CAMBRIDGE LIBRARY COLLECTION

Books of enduring scholarly value

Technology

The focus of this series is engineering, broadly construed. It covers technological innovation from a range of periods and cultures, but centres on the technological achievements of the industrial era in the West, particularly in the nineteenth century, as understood by their contemporaries. Infrastructure is one major focus, covering the building of railways and canals, bridges and tunnels, land drainage, the laying of submarine cables, and the construction of docks and lighthouses. Other key topics include developments in industrial and manufacturing fields such as mining technology, the production of iron and steel, the use of steam power, and chemical processes such as photography and textile dyes.

James Watt

This 1936 book, published to celebrate the bicentenary of Watt's birth, examines his career as a craftsman and engineer, rather than offering a purely narrative biography. Watt began his life as a maker of mathematical instruments, and throughout his working life enjoyed the challenge of such skilled work. Watt's inventions did much to power the Industrial Revolution and its economic and social consequences. However, he owed much of his commercial success to his long partnership with Matthew Boulton, a far more astute businessman, and a considerable portion of the book is devoted to the achievements of this period. An engineer by profession, H.W. Dickinson researched widely, and published highly readable works on the steam engine, Watt, Boulton and Trevithick. He succeeds in producing a work which appeals to the scientist, the historian and the general reader, without feeling obliged to over-simplify the technical details.

Cambridge University Press has long been a pioneer in the reissuing of out-of-print titles from its own backlist, producing digital reprints of books that are still sought after by scholars and students but could not be reprinted economically using traditional technology. The Cambridge Library Collection extends this activity to a wider range of books which are still of importance to researchers and professionals, either for the source material they contain, or as landmarks in the history of their academic discipline.

Drawing from the world-renowned collections in the Cambridge University Library, and guided by the advice of experts in each subject area, Cambridge University Press is using state-of-the-art scanning machines in its own Printing House to capture the content of each book selected for inclusion. The files are processed to give a consistently clear, crisp image, and the books finished to the high quality standard for which the Press is recognised around the world. The latest print-on-demand technology ensures that the books will remain available indefinitely, and that orders for single or multiple copies can quickly be supplied.

The Cambridge Library Collection will bring back to life books of enduring scholarly value (including out-of-copyright works originally issued by other publishers) across a wide range of disciplines in the humanities and social sciences and in science and technology.

James Watt

Craftsman and Engineer

H. W. DICKINSON

CAMBRIDGE UNIVERSITY PRESS

Cambridge, New York, Melbourne, Madrid, Cape Town, Singapore,
São Paolo, Delhi, Dubai, Tokyo, Mexico City

Published in the United States of America by Cambridge University Press, New York

www.cambridge.org
Information on this title: www.cambridge.org/9781108012232

© in this compilation Cambridge University Press 2010

This edition first published 1936
This digitally printed version 2010

ISBN 978-1-108-01223-2 Paperback

JAMES WATT

LONDON
Cambridge University Press
FETTER LANE

NEW YORK · TORONTO
BOMBAY · CALCUTTA · MADRAS
Macmillan

TOKYO
Maruzen Company Ltd

JAMES WATT'S WORKSHOP AT HEATHFIELD, *c.* 1895

Courtesy of J. H. Tangye, Esq.

JAMES WATT

CRAFTSMAN & ENGINEER

by

H. W. DICKINSON

Joint Author of
'James Watt and the Steam Engine',
'Richard Trevithick, the Engineer and the Man'

CAMBRIDGE
AT THE UNIVERSITY PRESS
1936

CONTENTS

Industry in England in the sixteenth to the eighteenth
centuries: Inventions. Improvements in transport. Act of
Union, 1707. Industry in Scotland, particularly in the Clyde
area. The craftsman. The gild system and its decay. Birth of
the engineer.

Boyhood, schooling and apprenticeship of Watt. Starts
business. Opens shops in Glasgow. Marries Margaret Miller.
Experiments with steam.

Repair of the atmospheric engine model. Separate condenser
engine. Dr John Roebuck interested. Meeting with Matthew
Boulton. Experimental work.

Surveys and reports on canals and other works in the
counties of Lanark, Argyll, Fife, Renfrew, Stirling and
Inverness. Improvements in surveying instruments. Bankruptcy of Roebuck. Boulton acquires his interest in the
condenser patent. Death of his wife. Removal to Birmingham.

PREFACE

THE life of a great man nearly always repays study from more than one point of view, and the greater the man, the greater the number of points of view that can profitably engage attention.

In the case of James Watt, the steam-engine, of which, by the way, he is erroneously credited by ordinary folk as being the inventor, is the one aspect that is always treated. There are other aspects of Watt about which much can be said and one of these is his craftsmanship. He was a craftsman at the outset of his career, craftsmanship helped him throughout his working life as inventor and engineer, and craftsmanship was the solace of his old age. No excuse seems necessary, especially as the bicentenary of his birth occurs next year, for presenting a study of Watt with this aspect in view, not neglecting of course the fundamental work that he did in the development of the steam-engine, amounting as it did almost to a re-creation.

The author is grateful to the Syndics of the Cambridge University Press for affording him the opportunity of treating James Watt's life from this point of view; it is an opportunity that is greatly appreciated because the author has been a student of the life and work of Watt for more than a quarter of a century. He collaborated with Mr Rhys Jenkins in the preparation of *James Watt and the Steam Engine*, the Memorial Volume issued in 1927 to commemorate the centenary of Watt's death. In so doing, the ground traversed was the same as that of the present work but, as frequently happens, the authors accumulated more material than they were able to use. Some parts of that material, particularly germane to the present work, have been incorporated with Mr Jenkins's approbation and help. Incidentally it should be remarked that the life of Watt is extraordinarily well documented—so that the task has been largely one of selection. Further, the author has striven to present the story in the very words of the protagonists, wherever space has permitted such a course. Some regret may be felt that references to authorities quoted are not given, but this would have overloaded the work. The author can only beg the reader to refer to the Bibliography at the end, and to the large work of Mr Jenkins and himself cited above. Here it may be well to state, to save the well-informed reader going farther, that he will find that no new conclusions about Watt and his work have been arrived at.

The Memorial Volume alluded to was issued by the Oxford University Press and the Delegates have been most obliging in granting permission to use material contained in it.

The author thanks the Director and Officers of the Science Museum, South Kensington, where are preserved the Watt engines, the Watt models and the Watt Workshop, for much material from those sources, as well as from Museum publications. Among the latter must be singled out *Science Museum Technical Pamphlet No.* 1, the subject of which is *The garret workshop of James Watt*, because great use of it has been made in the last chapter of the present work. To do so the permission of the

Controller of H.M. Stationery Office has been readily given and is hereby gratefully acknowledged. The Council of the Institution of Mechanical Engineers instantly gave permission for extensive use to be made of a paper that the author presented to that Institution in 1915, and this permission is highly appreciated.

Mrs Gibson Watt of Doldowlod, the present representative of the Watt family, has graciously given the author authority to use extracts from the documentary matter preserved there.

Mr Arthur Westwood, the Assay Master of Birmingham, in whose office the Boulton papers are now preserved, has been of material assistance in verifying transcripts and supplying photostats of letters which the author copied when the documents were still in the possession of the Boulton family at their seat, Tew Park, Oxon. The Chief Librarian of Birmingham, Mr H. M. Cashmore, is thanked for help with material from the Boulton and Watt Collection, now in the Reference Library; similarly the Director of the Birmingham Art Gallery, Mr S. C. Kaines Smith, is thanked for the permission to use the portrait of Murdock.

It gives the author abiding satisfaction to be able to say that his son, H. D. Dickinson, M.A., has read through the MS. and has given valuable help. Many other good friends have helped the author and he thanks them collectively since they prefer to remain anonymous.

H. W. D.

July 1935

LIST OF PLATES

WOODCUTS AS TAIL-PIECES

LIST OF FIGURES IN THE TEXT

CHRONICLE

of the

LIFE & WORKS OF JAMES WATT

Craftsman and Engineer

Age	Year	
	1736	Born at Greenock, January 19th.
18	1754	Begins life in Glasgow.
19	1755	Spends year in London mathematical instrument-making.
21	1757	Opens shop in the College of Glasgow.
23	1759	Enters partnership with John Craig and opens shop in the Saltmarket.
27	1763	Repairs Newcomen engine model belonging to the College. Removes to shop in the Trongate.
28	1764	Marries his cousin, Margaret Miller.
29	1765	Death of Craig. Conceives the separate condenser steam-engine.
31	1767	Takes up land surveying. Dr John Roebuck of Carron takes an interest in Watt's invention. Journeys to London on canal business.
32	1768	First meeting with Matthew Boulton.
33	1769	Takes out patent for separate condenser engine and erects one at Kinneil.
37	1773	Roebuck in financial straits. Death of Mrs Watt.
38	1774	Boulton takes over Roebuck's share in the patent. Removes to Birmingham, resumes experimental work on the engine and attains success. Joins the Lunar Society.
39	1775	Patent extended for twenty-five years, and partnership with Boulton for like term begun.
40	1776	Constructs pioneer engines at Bloomfield Colliery and New Willey Ironworks. Revisits Scotland and marries Ann MacGregor.

Age	Year	
41	1777	Journeys to Cornwall and erects his first engine there.
		William Murdock enters the service of Boulton & Watt.
42	1778	Murdock sent to Cornwall.
44	1780	Patents letter-copying process.
45	1781	Jabez Hornblower patents compound engine.
		Patents substitutes for crank.
46	1782	Patents expansive working, double-acting engine and rotative engine.
		Reasons that water is a compound body.
		John Southern enters the service of Boulton & Watt.
47	1783	Constructs first rotative engine.
48	1784	Patents parallel motion, steam carriage, etc.
		Constructs first double-acting engine.
		Elected Fellow of Royal Society of Edinburgh.
49	1785	Patents smoke-consuming furnace.
		Elected Fellow of the Royal Society, London.
50	1786	Visits Paris with Boulton.
51	1787	Introduces bleaching by chlorine from France.
52	1788	Applies governor to steam-engines.
54	1790	Buys land at Handsworth and builds Heathfield.
55	1791	"Church and King" riots occur in Birmingham.
57	1793	Litigation against infringers of first patent begins.
58	1794	Firm of Boulton, Watt & Sons established.
59	1795	Soho Foundry built.
61	1797	Studies medical chemistry.
62	1798	Buys landed estate at Doldowlod, Radnor.
63	1799	Validity of first patent established.
64	1800	Expiration of first patent and termination of original partnership.
68	1804	Commences work on sculpturing machines.
73	1809	Death of Boulton.
78	1814	Elected Foreign Associate of the French Academy.
79	1815	Death of Southern.
83	1819	Dies at Heathfield, August 25th.

CHAPTER I

INTRODUCTORY

Industry in England in the sixteenth to the eighteenth centuries: Inventions. Improvements in transport. Act of Union, 1707. Industry in Scotland, particularly in the Clyde area. The craftsman. The gild system and its decay. Birth of the engineer.

IN order to understand the situation in Great Britain in the middle of the eighteenth century when the story unfolded in the present volume begins, it is desirable to know somewhat about foregone events and the forces that moulded them. A brief review of the facts bearing immediately on our subject matter is all that is possible here; deeper knowledge of the political and social background, although desirable even for this small work, must be sought elsewhere.

Fortunately, for our purposes, it is unnecessary to go further back than the Tudor period; up to that time England had been well known as an agricultural and mining country producing raw materials, and self-supporting in the matter of food and clothing. It obtained luxury and other products largely by the export of wool, but also by that of other materials such as the tin of Cornwall which was known all over the world. The closest parallel to medieval England that we can find at the present day is Australia.

The policy of the Tudor sovereigns was to foster trade and commerce and under statutory protection to introduce new manufactures and industries or develop

older ones; in this way for example the art of deep
mining, the blast furnace, glass making and paper
manufacture were introduced during the sixteenth
century. The overseas trading to which these activities
gave rise led to the introduction of shipbuilding and
concomitant trades. In the main these advances were
effected by the introduction of skilled artisans from the
continent of Europe.

As a consequence the population of England, hitherto
only about five million souls, began to increase, towns
grew in size or new ones sprang up, and a middle class
emerged, engaged in trading and the direction of industry. A ferment of new ideas, evidencing itself in a
critical attitude towards current religious belief and
practice, brought about the Reformation. The religious
houses, owing to the acquisitiveness of land and pos-
sessions that they displayed, and owing to the conduct
of their members, had outlived much of their hold upon
the common people, hence the small resistance that was
offered when Henry VIII proceeded in 1536 to dissolve
the monasteries. The bulk of the domains of these
communities, which had been added to very extensively
during the preceding centuries by pious benefactors,
passed into the private ownership of court favourites
and others. The economy of these domains, static for
so long, was rudely broken down and this facilitated a
change to a newer economy. In the main the change
was from a subsistence agriculture to one of production
of raw material for industry—cornfields were turned
into sheep walks. The search for metals was prosecuted

with vigour and mines were opened out. The consequent upheaval in the incidence of employment coupled with the miserable condition of the former eleemosynaries of the monasteries brought about distress and vagrancy: to meet it the first Poor Laws, subsequently consolidated in Queen Elizabeth's reign, were passed.

Intolerance of the old for the new led to religious persecution both in this country and abroad. Waves of immigration arising from this cause, such as the persecution of the Spanish Inquisition in the Netherlands, 1568, and the massacre of St Bartholomew, 1572, were received on our shores. Such immigrants, if not welcomed, were not refused admission or repatriated; since they were largely skilled artisans they increased the pace of, even if they did not add to, technological progress.

The prevailing feeling was that of awakening after long sleep to fresh life and hope. A spirit of adventure was abroad: voyages of discovery both east and west, again furthered by Tudor policy, opened out new vistas of wealth to be acquired by conquest or by commerce. The opening up of the New World tended to shift the centre of gravity of commerce further west, nearer to the new countries. This increased the importance of the British Isles among other countries, and lessened the importance of those on the Mediterranean seaboard and with it that of her great republican cities such as Venice. In a word the age of individualism, capitalism and materialism was insensibly being ushered in. The mariner's compass, the art of printing and gunpowder were the outstanding inventions that assisted this new orientation of men's activities.

The seventeenth century saw increasingly rapid political and economic changes and was marked particularly by intellectual activity. The mind of the Englishman of the period was vivacious, inquisitive, speculative and revolutionary. The *nouveaux riches* landlords actively welcomed material improvements such as sea and river embanking, fen drainage, water supply, deep mining and new systems of agriculture.

There was a rage for establishing monopolies. The grant of one was the method by which the sovereign rewarded a favourite. It was not as if the monopoly was conditional upon the rendering of some service such as bringing into the country a new industry, as had been practised in earlier reigns; on the contrary, the monopoly was one for the exclusive import, control or manufacture of some article of general use or consumption such as soap, in fact such a monopoly as could be expected to bring in a large return. The purpose of the grant so far as the sovereign was concerned was to bring in revenue for himself without having recourse to Parliament; the grant was therefore conditional upon a substantial share, or fixed payment being reserved as revenue for the Crown. This led to serious abuses, so great indeed that the Statute of Monopolies, 1624, was passed to stop it. Monopolies were thereby limited to first and true inventions, and the statute laid the foundation of our patent system which with all its drawbacks has been a potent engine for fostering invention.

At the end of the seventeenth century, as a result of the revocation of the Edict of Nantes by Louis XIV, in 1685, a further wave of immigration, this time of French Huguenots, entered this country. Many of them were skilled in weaving and other trades, others were merchants or traders; on the whole they were a great asset to the country of their adoption.

In the political sphere, the struggle against a strong monarchy in the person of Charles I culminated in the two civil wars, his execution and the short-lived Commonwealth. The restoration of the monarchy in the person of Charles II revived the political struggle in a modified form and the issue was not finally settled in favour of a limited monarchy till the abdication of James II in 1688. The Crown and the aristocracy were forced to share political power with the more successful members of the newly enriched class of improving landlords, merchants, stock jobbers and industrialists.

On the intellectual side the seventeenth century was of marked importance. The barren speculations of the schoolmen had given way to the conviction that the approach to knowledge must be, as laid down by Francis Bacon and others, by way of experiment, observation and inference. The renaissance of learning fostered by intercourse with other civilisations, existing and past, brought with it much that was valuable.

This led in all parts of the continent of Europe to the foundation of academies and scientific societies such as our own Royal Society in 1663. This was the beginning of scientific research; the soil was virgin, labour and study were richly repaid, yet the field was still sufficiently circumscribed for a vigorous mind to keep in touch with all that was going on. To this century we owe the knowledge of the law of gravitation, the laws of gases, the knowledge that the atmosphere has weight, and the invention of the air pump.

Decidedly it was the aim of men of science to be of practical use in directing the forces of nature. The technical men, however, carried out their work in the traditional methods of their predecessors and seem to have owed little to the new learning. They led rather than followed: for example, Galileo's and Torricelli's attention was drawn to the fact that the atmosphere had weight, and to the invention of the barometer, because the makers of pumps found they could not draw water from a depth of more than about 28 feet. Demands on an ever-increasing scale for such practical needs as the supply of water to towns were met by the multiplication, often on a gigantic scale like the Machine of Marly near Paris, of existing contrivances —pumps, water wheels, gearing and link-work. Engineering was still empirical. The spread of information was being assisted by the publication newly begun on the continent of works on mining and mechanics of a more or less encyclopaedic character. How great was the value of these books will be seen later.

The eighteenth century was characterised in England by a great outburst of technical invention and by the

rise of industry on a really large scale—a phase which
has been designated the industrial revolution. Many
factors contributed to this end. Banking by private
individuals had slowly taken root in Great Britain and
the formation of the Bank of England in 1694 and the
Bank of Scotland in the succeeding year may be said
to mark the beginnings of modern banking. The rise
of chartered companies on a joint stock basis stimulated
enterprise not only in newly-acquired lands such as
India (the Honourable East India Company, 1600),
but also at home (the South Sea scheme, 1711). Some
of these companies turned from their avowed objects
to engage in land-owning and finance. Speculation
became rife, all classes of society found excitement in
gambling in the shares of companies promoted for the
wildest possible schemes, and this culminated in the
frenzy of the South Sea Bubble of 1720; the ruin
wrought was widespread.

The outstanding invention of the eighteenth century
was that of the steam-engine, due to the genius of
Thomas Newcomen of Dartmouth in Devon in the
year 1712. Ever since man had advanced from a mode
of life, in common with that of other animals, depending
solely on the products to be obtained on the surface of
the earth, and had sought the superior command over
the forces of nature afforded by the possession of metals,
he had been confronted with the difficulty of the
gradual exhaustion of minerals found on the surface;
this led to the prosecution of the search for them below
ground. Mining by open working or by adit was first

practised and was comparatively simple, but as the search went deeper, that great enemy of the miner—water—was encountered. Every means of pumping by animal and water power was used, but these were inefficient and ineffective for all but the shallower mines—a limit seemed to be set to further advance. The need called forth, as it so frequently does, the remedy: the harnessing of the power of steam.

Not only did the steam-engine solve the problem of unwatering mines to greater depths than was possible with animal power, but it began, almost unsuspectedly, a new chapter in the history of civilisation. For the first half century of the existence of the steam-engine, its sole application was to raising water. It was not yet applied to industry generally; the needs of industry, such as they were, were met by the utilisation of the water powers which the climate and configuration of these islands render available, on a small scale, throughout the year, without excessive outlay for impounding or damming. It is with the development of the steam-engine at the hands of James Watt to fulfil this far greater rôle of power supply generally that the present volume is concerned.

Another development of the highest importance for Great Britain was the solution by Abraham Darby about 1713 of the problem of smelting iron by mineral fuel, thereby releasing the industry from its previous dependence on the charcoal obtainable from our all too exiguous woodlands. It is not generally realised that in 1740 the annual output of iron in England was only

17,000 tons. Apart from the advantage of being able to rely upon the mineral fuel with which this country is so well supplied, there was an advantage of a more positive kind, viz. that iron could be procured so plentifully and cheaply that its use could be widely extended, particularly in the form of castings, thus providing a new structural material of very high importance. Wrought iron also became more abundant as the direct but wasteful process of the bloomery had been superseded, i.e. about the time of the introduction of the blast furnace, by the indirect process of obtaining it from cast iron by the agency of the refinery hearth. Iron could now be made in sufficiently large masses to be forged under hammers driven by water power into plates and bars of sufficient size for boilers and other structures to be made from it.

Consequent upon these developments arose a demand for improved means of communication and of transport of goods. This, in these hilly islands, stimulated coastwise trade; fishing villages blossomed out into seaports, piers and harbours were built, docks were made and dredging was begun. Nearly every port in the United Kingdom developed ancillary industries such as shipbuilding and rope-making. Further, since inland communication was inhibited by bad roads, improvements were made by the canalisation of rivers; this led in turn to the making of locked canals or navigations, as already exemplified in France, Holland and elsewhere. We need only mention as prominent in this enterprise the Duke of Bridgewater and James Brindley whose canal from Worsley to Manchester was finished in 1760.

An important political event was the Act of Union of the kingdoms of England and Scotland, which was effected in 1707. It had far-reaching effects, particularly in Scotland which emerged almost at a bound from a state that was little removed from a feudal and tribal one into the new economic order of her neighbour.

Scotland at the beginning of the eighteenth century, as a whole, had scarcely any industries. An example or two must suffice. Coal mining, as distinct from outcrop working, had begun in Fifeshire and in the Lothians where the atmospheric engine had been introduced. Attempts to introduce the charcoal blast furnace using rich hematite ore from Furness, between 1727 and 1736, met with failure. The smelting with mineral fuel of the local low grade clayband ironstone, of which there were vast quantities, was not effected till the Carron Ironworks were founded in 1760 by the enterprise of Dr John Roebuck and his associates.

The east of Scotland had trade with the North Sea ports and the Baltic, hence towns like Aberdeen had risen to prominence, shipbuilding had developed to some extent, and a seafaring population had grown up. The west of Scotland, however, was almost destitute of industries except perhaps for fishing and fish-curing. The closest parallel that we can draw is with the Orkneys and Shetlands of to-day. There were few towns of any importance in the west beyond the royal

burghs of Ayr, Renfrew, Paisley and Glasgow. The
latter, for example, was a sleepy cathedral and university city, clustered along the banks of the Molendinar,
a tributary of the River Clyde. Glasgow was not yet a
port, for the Clyde meandered thence through a shallow
estuary for about fifteen miles before it reached deep
water. No boat larger than 10 tons' burden could come
up to the Broomielaw. Indeed, the estuary of the Clyde
had little more significance than that of a ditch, guarded
by the stronghold of Dumbarton Castle, separating the
turbulent clans of the Highlands from their slightly
more peaceable neighbours of the Lowlands.

The idea of deepening the estuary of the Clyde had
not yet been conceived, but the absence of water transport had been partly remedied by the founding by the
magistrates of Glasgow in 1668 of the town of Port
Glasgow, twenty miles west-north-west of the City. In
1710 it was the main customs' house on the Clyde. There
were already other places on the estuary more favourably situated for deep water and anchorage such as
Greenock which, raised to the dignity of a burgh of
barony in 1641, had entered upon a period of activity
by the construction of the East and West Quays, begun
in 1706, to form a harbour, the largest undertaking of
its kind in Scotland at that time.

Another but smaller town was Crawfordsdyke or
Cartsdyke, a near neighbour of Greenock, created a
burgh of barony in 1669, and eventually swallowed up
by the latter. Both of them shared in the provision trade
with Ireland, and in that of tobacco leaf with Virginia

and Maryland, then springing up. Shipbuilding followed and Cartsdyke had the distinction in 1718 of
building the first ship that sailed from the Clyde in the
tobacco trade. The first square-rigged vessel was not
built till 1768—this was at Greenock—and even till late
in the century the vessels built there averaged under
100 tons' burden. Larger vessels were employed, it is
true, but they were built in New England. The population of Greenock in 1741 was only 3381 souls, while that
of Cartsdyke was only 719.

Industry during the period under review was carried
on by the workers themselves without the aid of an
entrepreneur, except in such cases as iron-making where
the plant was large and costly; in such cases the landlord
provided the plant and materials for the workers who
paid him on a contributory basis. A few words should
be said about the position of the worker—the skilled
craftsman. There had been a constant effort on his part
to retain the position that he had gained in the Middle
Ages, when in every town of any importance those
practising a particular craft had associated themselves
to form a gild, such as those of the bakers, weavers,
fullers, dyers, salters, fishmongers, goldsmiths and
leather-sellers.

On the whole, the influence of the new industrial
order on the gild principle was a disintegrating one.
Besides, new towns such as Birmingham, that knew not
gilds, sprang up and by the very absence of restrictions
attracted new industries and gildless workers at the
expense of neighbours like Coventry where the gilds

were still powerful. Even within the corporate town, and within the gild itself, decay had set in for the gilds became oligarchic and conservative. In the sixteenth century the merchant master employing small working masters each with his complement of journeymen grew up. The latter had no likelihood of ever becoming a master; he thus lost control of his own working conditions and sank to the position of a mere wage-slave. In most towns the gilds simply died out; in a few places only such as London they survived but in the much modified form we see to-day, with few obligations and few duties, yet with privileges in the election of the Corporation. One can be a Leather-seller for example without knowing how leather is made. Within the last year in London we have seen the incorporation of three fresh city companies of this atrophied kind.

Certain crafts, e.g. that of the barber, have developed into professions and somewhere in this category we must place the engineer. In its origin the occupation was military—that of the artificer who worked the ballista, catapult or "gin" used in attacking walled cities—the "ingyner", or "engynier" (latinised as "ingeniator", and shortened to "Ginner" or "Jenner", whence these two surnames). As the art of war increased in complexity and the place of the ballista was taken by the cannon, the engineer's duties were enlarged and his importance increased. This is not the place to trace his upward progress. Suffice it to say that when in the eighteenth century a word was needed to describe the work of such men as were directing the forces of nature for the use and convenience of man and not for his destruction, the name was ready to hand and only needed the addition of the word "civil" to differentiate his employment from that of his military counterpart.

The early civil engineers arose from the most various occupations. Sir Cornelius Vermuyden, who drained the Lincolnshire fens in the time of Charles I, was trained in sea embanking in Holland. Sir Hugh Myddelton, who brought the New River to London, was a goldsmith. John Perry, who repaired Dagenham Breach, was a naval officer. Pierre Paul Riquet, the constructor of the Grand Canal of Languedoc, was a tax-gatherer. James Brindley, already mentioned, was a millwright. John Rennie was another and the erectors of the early steam-engines were drawn largely from this trade. John Smeaton started life as a mathematical instrument maker and so did James Watt, as we shall see presently. And so we might go on citing other instances of masons, blacksmiths, carpenters, land surveyors, colliery viewers, and others, who have adorned the profession of engineering with their talents.

Even to-day, when the amount of knowledge to be acquired is so great and the years of training to be undergone are so many before a man is qualified to be an engineer, he must, in this country at any rate, have had practical experience in the field or in the workshop before he is considered fully trained. Thus the engineer is a craftsman and nearly everything that we see around us in the material world is rooted in craftsmanship.

CHAPTER II

BEGINNING OF CAREER,
1736–1763

Boyhood, schooling and apprenticeship of Watt. Starts business. Opens shops in Glasgow. Marries Margaret Miller. Experiments with steam.

IT was into such a world, and such a part of it as we have attempted to picture in the previous chapter, that James Watt was born on January 19th, 1736, in the small seaport town of Greenock in Renfrewshire. The Register of Baptisms, 1736, of the Old Parish records the event succinctly thus:

> WATT, James Son law'll to James Watt Wright in Gr. and Agnes Muireheid his Spouse was born the 19th and baptised the 25th.

The word "lawfull" points to a whole chapter of social history!

There is but little known about the forbears of Watt. His grandfather Thomas Watt (1642–1734) had come, late in the seventeenth century, from Aberdeenshire and had settled in the burgh of Cartsdyke, which, as we have stated, was a near neighbour of Greenock.

Thomas Watt was living in Cartsdyke as early as 1683; he is described usually as a "mathematician"; in the Register of Burials he is styled "teacher of navigation" and that is what his occupation really was. It may be that in coming south he looked upon Cartsdyke and Greenock as rising towns that would afford scope for his talents in teaching navigation to such mariners as wished to become masters. In the mathematics course at the Marischal College, Aberdeen, navigation was included and rightly so, for that city was then the chief port of Scotland. A Thomas Watt is recorded as having entered this College in 1665 but no further mention of him occurs. The reader may put two and two together for himself.

Thomas Watt rose to a position of some importance in Cartsdyke, being appointed as early as 1696 Bailie of the Barony, an official who in the then semi-feudal order exercised the direction of local affairs under his superior lord. Thomas had six children: the eldest surviving son, John, became a surveyor in Glasgow; a map of the Clyde estuary survives to furnish evidence of his abilities. The second surviving son, James (1698–1782), was apprenticed to the trade of wright or, as we south of the Tweed would say, that of carpenter. He prospered with the growing prosperity of the town and, as frequently happens in a small place, he carried on other businesses. He was a builder, not only of houses but of ships, a ship-chandler, and, later, a merchant; in other words, he acquired interests in all the businesses that were going on in the town. Like his father before him he attained to official positions there, e.g. in 1755 he was elected Treasurer and in 1757 Bailie of the Burgh. He married in 1729 Agnes Muireheid or Muirhead, a woman of superior intellect and force of character who, as was then thought proper, confined her talents to the household. There was a strain of bodily weakness somewhere in the family, probably on the father's side, for their first five children died in infancy, so that it is not a matter for surprise that Jamie, the eldest surviving child, should have been delicate. He must have required more than ordinary care in upbringing, and it is only natural that his mother's affection should have been centred on him, possibly to the extent of spoiling him somewhat. His childhood was uneventful except for the excitement during the "Forty-five" when Bonnie Prince Charlie was believed to have landed at Greenock and a house-to-house search was made for him.

His mother taught Jamie his letters and in course of time he was sent to school where, as one would expect in the case of a delicate and thoughtful boy, he was bullied, and this to such an extent that he shone neither in lessons nor in the rough games of his schoolmates, in fact he was considered "dull and inapt". He passed on to the Grammar School, where he underwent the usual classical curriculum with some advantage to his mental powers and to his knowledge of Latin. It was not till he was about thirteen and rose into a mathematics class, taught by one John Marr, that his ability began to show itself.

George Williamson, the biographer of Watt's early years, who gathered his information from those who knew the family, gives a telling picture of the lad's environment while he was still at school:

The reader is already aware of the general nature of the father's business at Greenock,—the worthy master-wright, merchant, bailie and treasurer of the town,—

and of its miscellaneous character. A glance here at the work-benches will give us a more accurate idea of the kind of work going forward. In addition to most of the minor details of carpentry, such as the outfit and supply of the shipping demanded, we observe the carving of ships' figure-heads, the making of gun carriages, of blocks, pumps, capstans, dead-eyes &c. &c. The "touching" of ships' compasses also is done here, and the adjusting and repairing of such nautical instruments as are yet in use. We notice further, among other things interesting to us, a piece of mechanical work which was the construction of the same enterprising and ingenious man,—the first *Crane* made at Greenock, for the convenience of "the Virginia tobacco ships" then frequenting the harbour.

What a fine school for James's "innate love and liking for handicraft"! It is not sufficiently realised what an influence contact with handicraft in early years has in directing and moulding later activities. The writer can testify to it in his own case, for even in his youth it was still possible to see all kinds of trades being carried on in one small town; in these modern days of specialisation and concentration of industry in factories, such contact is scarcely possible.

After leaving the Grammar School, young Watt continued to occupy himself in his father's workshop, but not, so far as we can learn, as an apprentice; his parents may have thought him not constitutionally strong enough to do so. Probably his father intended Jamie to follow him in his business of merchant, or possibly his mother hoped in a year or two to send him to the College in Glasgow to fit him for some professional career, always a characteristic desire of a Scots mother. Williamson says that "he had here [i.e. in his father's shop] a small forge erected for his particular use" and we are to think of him "busy with chisels and tools" making models and acquiring that first-hand knowledge of materials that was afterwards to stand him in such good stead. The workmen used to say that "Jamie has gotten a fortune at his fingers' ends".

Williamson goes on to retail a story told him by one of the apprentices in the Watt workshop, who "remembered having been sent, when a boy, to clear out an attic room in his employer's house, where he found a quantity of...ingenious models...which Mr Watt senior told him had been, some years before, made by James, who was then in business in Glasgow. Among these models he remembered, in particular, a miniature *Crane*, and a *Barrel-organ*." We shall find how persistent in his after life was this early formed practice of making models. Williamson suggests that young Watt took up such light work as repairing the nautical instruments left with his father for that purpose; if so, this is evidence of the bent that he showed later.

His mother's death in 1753, and some reverses that his father experienced in business about the same time, made it necessary for the lad to decide what he was going to do for a living. He came to the conclusion that he wanted to be a mathematical instrument-maker, so in June 1754 he packed up his tools, his leathern apron and his small wardrobe, and was sent to his mother's kinsfolk in Glasgow to see what they could do to help him. Glasgow was, as we have said, a cathedral and university city, as yet only beginning that industrial expansion for which later it was to be so widely known. It will not altogether surprise the reader, therefore, to learn that there was no one carrying on the trade of mathematical instrument-making in the royal burgh. The best that could be done was to put Watt under an optician who was little better than a mechanic. Watt remained in Glasgow till the following May but, while deriving much advantage from the intellectual circle to which his relatives introduced him, he realised that he was wasting his time so far as learning a trade. Through his kinsman, George Muirhead, who held the chair of Humanity in the College, Watt got an introduction to Professor Robert Dick who, sympathising with Watt's aspirations and knowing that nothing of the desired craft was to be learnt in Glasgow, advised him to go to London, which was then the mecca of English instrument makers, for it was there that men like Ramsden, who had the highest reputation in the world, were established. Watt may have been spurred on by the idea that there would be a good opening later for him to set up in business as a mathematical instrument-maker in Glasgow, while Professor Dick may have thought it would be to the advantage of the College to have a man of that trade close at hand. Professor Dick did more, he gave Watt an introduction to James Short, a Scotsman, in that business in the Strand, in London.

Young Watt returned to Greenock to consult with

his father about the proposal, the decision was made, and on June 7th, 1755, with two guineas in his pocket and his father's blessing, he set out for the metropolis in company with Ensign John Marr, son of his old schoolmaster. There was no coach running then and the journey was made on horseback, travelling by Coldstream, Berwick, Newcastle-on-Tyne, and the great North Road. The two young men spent twelve days on the journey, for they would not travel on the Sabbath. Slow as the journey was, horseback was faster than the stage waggon, the alternative means of conveyance, but it meant if the traveller was not coming back at once that he had to buy his horse at the start and sell it at the end of the journey. Watt's chest of belongings was sent by sea from Leith to London; he might of course have gone that way too, but there were the risks of shipwreck and of the king's enemies.

Arrived in town, Watt's immediate business was to find a master, no easy matter, for he had served no apprenticeship and he was now too old for it. He could not rank as a journeyman because he had not worked under a master; even if he had done so in one town, that did not mean necessarily that he would be allowed to do so in another; in fact, so far as the gilds were concerned, he was a "foreigner". In July he wrote to his father: "I have not yet got a master, they all make some objection or other" and no wonder, for who wanted such an "offcome"? Mr Short did not take him, no doubt for some good reason, but sent him, after many rebuffs elsewhere, to John Morgan who had a shop in

Finch Lane, Cornhill, who was induced by an offer of twenty guineas and Watt's services to allow the latter to work there for one year. The lad wrote to his father: "If it had not been for Mr Short I could not have got a man in London that would have undertaken to teach me, as I now find there are not above five or six that could have taught me all I wanted.... Though he works chiefly in the brass way, yet he can teach me most branches of the business, such as rules, scales, quadrants, &c." Watt was allotted the least eligible bench in the shop—that nearest the door and in the draughts—but full of determination he set to work; although he was well grounded, he was attempting the well-nigh impossible feat of crowding into one year work that normally required three or four.

Watt's assiduity and progress were really remarkable, as we find recorded in letters he wrote home. By the 5th of August he was advanced to work on a Hadley quadrant and by the 23rd he had completed such a one better than a fellow apprentice of two years' standing. In October he had started on rules, by November on azimuth compasses. "We work to nine o'clock every night except Saturdays" and if that was not enough he durst hardly stir abroad for fear of being taken by the press gang to serve in his Majesty's Navy or, worse still, by kidnappers to work on the plantations. The press was occasioned by the Seven Years' War with France (1755–1762) for which English ministers were found ill-prepared, and hence the hasty measures to man our ships of war. If a man were pressed he might possibly

return home, but if kidnapped, never, for it was in effect
a sentence of death; we have to remember that those
were the good old days! In March 1756 Watt wrote:

They now press any one they can get, landsmen as
well as seamen, except it be in the liberties of the City,
where they are obliged to carry them before my Lord
Mayor first, and unless one be either a 'prentice or a
creditable tradesman, there is scarce any getting off
again. And if I was carried before my Lord Mayor,
I durst not avow I wrought in the City, it being against
their laws for any unfreeman to work, even as a journeyman, within the Liberties.

With the spring the close confinement and intense
application, for Watt had not previously experienced
either, became almost unbearable. He longed for the
air of his native "hills and the sea", but he held on
indomitably. In April he was able to say proudly:
"I think I shall be able to get my bread anywhere, as
I am now able to work as well as most journeymen,
though I am not so quick as many." In June he could
"make a brass sector with a French joint, which is
reckoned as nice a piece of framing as is in the trade".
He lived on eight shillings a week so as to be as small a
charge as possible upon his father; he could not live on
less, without, he said, "pinching his belly". The confinement, the long hours, the poor food had told upon
him; he had never worked so hard before. He had a
"racking cough, a gnawing pain in the back and weari-
ness all over the body". Already doubtless he suffered
from the sick headaches and consequent depression
which were his companions far beyond middle age.

In July, at the expiration of the year for which he had bargained, he said good-bye to his master, and with £20 worth of materials and tools that he would need wherewith to set up in business in Scotland, and a copy of Bion's *Construction and use of Mathematical Instruments*, 1723 (Edward Stone's translation), he set off home by the way he had come, proud to reckon himself a full-blown craftsman.

He found his father, if not too prosperous, yet high in the estimation of his fellow-townsmen, by whom he was advanced to the dignity of bailie and likewise elder of the kirk. Young Watt, now in his twenty-first year, stayed at home till October recruiting his strength and then went to Glasgow to see Dr Dick, only to find him expecting the arrival of some astronomical instruments bequeathed to the College by Alexander Macfarlane, a merchant of Jamaica. They arrived while Watt was there and as several had "suffered by the sea air, especially those made of iron" during the voyage, he was desired "to stay some time in town to clean them and put them in the best order for preserving them from being spoiled", a job he was only too glad to undertake. To enable him to do so, he was given a room in the College looking into the quadrangle. It was then that Watt got to know the new professor of Anatomy and Chemistry, Joseph Black, and the young graduate John Robison, who were to prove life-long friends.

In December, Watt was paid £5 for this work, possibly the first money he had earned since his apprenticeship; he had apparently returned home and he may have had some idea of setting up business in Greenock, which at that time one would have said was as suitable a venue as Glasgow. This is suggested by the fact that his *Waste Book* contains "An Inventory of the Goods, Money, Debts &c. Belonging to me, James Watt Junr., also what I owe to others £21. 4. 2.", dated "Greenock, Jan. 3rd 1757". If he had entertained such an idea, he gave it up and determined to set up in business in Glasgow; this is shown by his *Waste Book*, in which on August 2nd he enters: "Expenses removing to Glasgow 7ˢ 8ᵈ." His first difficulty, of course, was to find a place of business. For one thing he possessed only about £20 of capital and for another thing he was not a burgess and, therefore, did not possess the right to set up as a craftsman within the city. With so many influential friends in the College, however, and the fact that he had been so recently accommodated there, it is not surprising that he was given permission to open a shop in the College itself—there were several precedents for such a course, e.g. the printers to the University were granted such a privilege—and to style himself "Mathematical-instrument-maker to the University". His *Waste Book* fixes the date as being December 6th, 1757.

Quite a story has been told—and retold with a snowball-like wealth of accretions—about the opening of this shop. How the Incorporation of Hammermen, the gild most cognate with instrument-making, persecuted Watt and refused him permission to set up in business in the city. There is no trace in the records of the Hammermen, or of the Dean of Guild Court which had jurisdiction in such matters, of any proceedings against Watt. The whole story rests on this plain statement by Dr Black: " Mr Watt came to settle in Glasgow as a maker of mathematical instruments; but being molested by some of the corporations who considered him an intruder on their privileges, the University protected him by giving him a shop within their precincts and by conferring on him the title of Mathematical instrument maker to the University." Watt certainly was an intruder and very probably he was told so in strong language by members of the gilds and perhaps by others. Being naturally timid, his instinct would be to run with these tales to the College where his friends were. It is to be remembered that the College was quite outside the jurisdiction of the city, and may indeed have been pleased to have had opportunities of marking the fact. This division of authority is one that was common in the Middle Ages and still survives among us, for example, at Oxford. The best proof that the story rests on a very shaky foundation is that Watt *did* set up a shop in the city only two years later.

The College (see Pl. II) was then situated, not in its present magnificent surroundings at Gilmorehill, but on the east side of High Street, leading up from Glasgow Cross to the Cathedral, on the site now occupied by College (Goods) Station of the London, Midland and Scottish Railway and of the High Street (Goods) Station of the London and North-Eastern Railway.

The space that was allotted to Watt in the College was on the ground floor of an old house "with a sort of

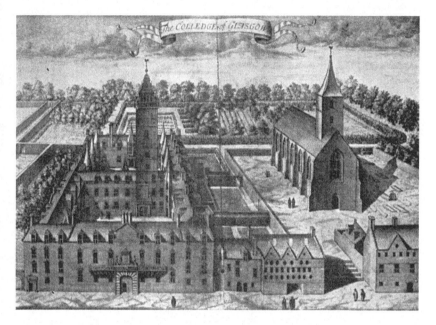

PLATE I. THE COLLEGE OF GLASGOW

From the engraving in Slezer's *Theatrum Scottiae*, 1693

arcade in front supported on pillars"; there is nothing in our illustration that we can identify as the spot.

In such pleasant surroundings, with congenial visits from professors and students, Watt would have been content if he had been able even to scrape a living, but such was not the case. Writing on September 15th, 1758, to his father he says: "Unless it be the Hadley's instruments, there is little to be got by it, as at most jobs I am obliged to do the most of them myself; and as it is impossible for one person to be expert at everything, they often cost me more than they should do." Incidentally this letter reveals the fact that he was employing workmen and that he realised the advantage to be derived from the division of labour.

But if he was not making money he was making friends, and among them we like to visualise Robison, three years the junior of Watt—"a philosopher as young as myself and always ready to instruct me"—lounging round the shop discussing all things under the sun, steam-engines and steam-carriages included, as young men just love to do.

Whether he made the proposed journey to Liverpool to find customers we do not know, but in August 1759 he made the intended journey to London to get orders for instruments. This time he did not sell his mount. His "expenses of journey to London" were £4. 6s. 3d. His "Expenses" there £5. 6s. 10½d. and "Expenses returning and keeping the mare there" £4. 18s. 6½d. We should like to know how successful he was in getting orders, and how long his stay was.

Probably he did succeed for evidently he now needed more capital; this was supplied by John Craig, said to have been an architect, with whom he entered into a partnership of equality. This began on October 7th, as shown by an *Inventory* still in existence; the tools were valued in round figures at £22 odd, the stock at £69 odd, and the cash in hand, including no doubt Craig's money, at £108. The stock comprised quadrant, gunter's scales, compasses, pencils, microscopes, temple frames, burning glasses, magic lanterns, etc., no doubt brought from London. The partners set up a shop in the Saltmarket, nearly opposite St Andrew's Street. The arrangement was that Watt received £35 a year as wages, and shared in the profits equally with his partner. Apparently business increased, for four years later he removed to a new shop in the Trongate, the principal street in the city, as this advertisement in the *Glasgow Journal* of December 1st, 1763, shows:

James Watt has removed his shop from the Sautmercat to Mr Buchanan's land on the Trongate where he sells all sorts of mathematical and musical instruments, with variety of Toys and other goods.

The mention of musical instruments is amusing, for Watt did not know one musical note from another, yet he mastered the theory of music in order to be able to make musical instruments; one at least of his wind organs is still in existence. It was made for St Andrew's Parish Church; and the story of the introduction of this "kist of whustles" into the kirk and of the opposition it encountered, as being ungodly, is most entertaining. "Toys" were the steel ornaments for which Birmingham was so well known and not what we mean by the term nowadays. Without doubt Watt was prospering, for he was employing several journeymen—no less than sixteen at the finish, and had even taken apprentices to "the instrument making".

He was extending his activities too, for in 1763 he acquired an interest in the Delftfield Pottery Company. In 1772 he had £474 invested in the concern. The pottery was situated on land south of Anderston Walk, now known as Argyll Street, extending to the banks of the Clyde at the Broomielaw. Through the works ran a lane, bounded by hawthorn hedges known as Delft-field Lane, widened and renamed long ago quite appropriately James Watt Street. It should be recalled that the manufacture of porcelain had only been introduced into England as recently as 1757 and experiments in many places were going on with the object of improving upon the rough delft ware then being made in Great Britain. At Delftfield we know definitely that in 1766 white stone ware that was harder in the "paste" than delft was introduced, in fact the pottery was the pioneer of this ware in Scotland. What share in this development may be attributed to Watt is not known, but that he was active we judge from the many allusions in his letters to experiments bearing on this subject. It is fairly safe to conclude that his work was at least of an advisory nature, such as the testing of clays, the layout of flint grinding mills and the construction of furnaces. It is from this employment that has arisen the statement, made by Josiah Wedgwood and others, that Watt once worked as a potter.

Watt in fact was anything but a dabbler in chemistry, for in 1765 he experimented for Dr Black and Dr Roebuck on a process of theirs for the manufacture of alkali by the decomposition of lime with sea-salt; the process did not come to anything, however. It may be that it was this experimental work that was the means of first introducing Watt to Roebuck. About this time, i.e. in 1765, Watt designed an apparatus for drawing in perspective (see Pl. II). A vertical board on which a sheet of paper is fixed is supported on three legs. Commanding the paper is a parallel ruler, in which is a socket for a lead pencil and which is provided with an index. An eyepiece is carried on a jointed arm fixed to the corner of the board. The object to be drawn is sighted through the eyepiece and followed by the index, the pencil meanwhile marking the lines on the paper. The board folds up to form a box of reasonable size for the pocket, and the legs telescope to form a staff about 4½ ft. long, which can be used as a walking stick. Watt made fifty to eighty of these apparatus—they cost three guineas—which were sent to all parts of the world. There are a number, partly finished, in the Watt workshop, to be described later. Our illustration is taken from one preserved in the Science Museum, South Kensington. Watt did not patent the apparatus with the result that it was copied, indeed one London maker, George Adams, put it in his catalogue of 1766 as if it were one of his own inventions and charged six guineas for it!

An important event in Watt's life was his marriage to his cousin, Margaret Miller. We know no details of the courtship, all we can say is that hers was a sweet

PLATE II. WATT'S PERSPECTIVE MACHINE, 1763
Courtesy of the Science Museum

nature and Peggy had great faith in Jamie. The wedding
took place on July 16th, 1764, and he brought her to
their new home in Delftfield Lane (see below)—he had
removed thither some time previously from the College,
although still retaining the shop there—and they started
housekeeping "on a very humble scale". Watt needed
someone at his back, for he was "modest, timid, easily
frightened by rubs and misgivings and too apt to de-
spond", and at no time did he need encouragement
more than now. He had begun to carry out experiments
on steam that proved to be of such importance that they
marked a turning-point in his life. Here, therefore, it
is convenient to pause in order to begin a new chapter.

Watt's home in Delftfield Lane, Glasgow, 1764
From Smiles's *Boulton and Watt*, 1865

CHAPTER III

THE SEPARATE CONDENSER,
1763–1769

Repair of the atmospheric engine model. Separate condenser engine.
Dr John Roebuck interested. Meeting with Matthew Boulton. Experimental work.

NO one could have imagined that, when Watt was
asked by Professor John Anderson during the
session of 1763–64 to repair a model of the
Newcomen or atmospheric steam-engine belonging to
the Natural Philosophy Class in the College, it was to
prove not only a turning-point in Watt's career but also
an important event in the history of civilisation.

The model had been sent to London to Jonathan
Sisson (*c.* 1690–1760), a well-known London instrument-maker, for repair but when it came back it was
found to be no better than before. Here was a job after
Watt's own heart. He knew a little about steam because
in 1761 or 1762 he had tried some experiments on high-
pressure steam with a Papin's digester, but he stopped
short because he realised the danger of bursting the
boiler and the loss of power if the vacuum were dis-
pensed with, as would be the case if the steam were to be
exhausted into the atmosphere; he had discussed, too,
with Robison, as we have already said, the idea of moving
wheeled carriages by high-pressure steam. Watt may be
said to have known nothing about the atmospheric

PLATE III. MODEL OF ATMOSPHERIC ENGINE REPAIRED BY WATT, 1765

Courtesy of Glasgow University

engine; examples in Scotland at that time could be counted on the fingers of two hands. As he says himself: "I set about repairing the model as a mere mechanician."

He quickly mastered its mode of action which, briefly, is this (see Fig. 1): a piston working in a cylinder is attached by a chain to a beam or lever rocking on trunnions, to the other end of which a pump rod is hung. Steam at atmospheric pressure, generated in a boiler below, is admitted into the cylinder and the air present is blown out through the snifting valve. The piston being overbalanced by the weight of the pump rod is at the top of its stroke. A jet of water is turned on in the cylinder to condense the steam and form a vacuum. The pressure of the atmosphere on the piston forces it down and in doing so it lifts the pump rod and with it water from the mine. With the readmission of steam, the cycle recommences. In practice the valves were opened and shut automatically by tappets in a plug or pump rod hung from the beam. Such was the apparatus that for half a century had been the only efficient means of draining mines or raising water for towns.

Watt found that the boiler of the model, although in proportion to the size of the cylinder, could not supply sufficient steam to keep the engine going for more than a few strokes. A model to scale has several inherent defects, since the area and volume vary as the square and cube of the linear dimension respectively, e.g. a vessel of half the linear size of another has only a quarter the heating surface and one-eighth of the volume. Also the effect of friction is much greater in proportion in a model than in a full-sized engine.

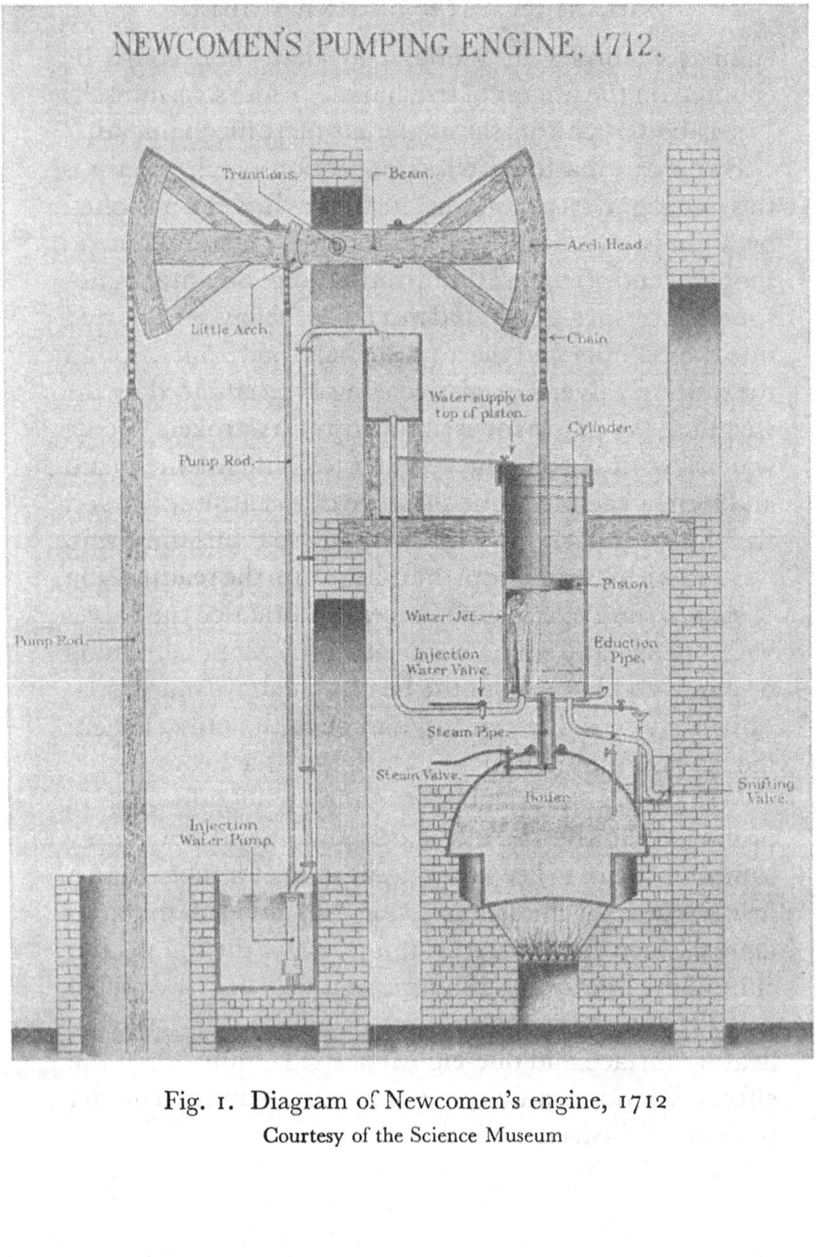

Fig. 1. Diagram of Newcomen's engine, 1712
Courtesy of the Science Museum

Watt knew from first principles that steam coming into contact with a cold body communicated heat to it and was condensed. He reasoned that, to avoid loss of steam by condensation when it entered the cylinder at the beginning of the stroke, the cylinder should be boiling hot, viz. 212° F.; on the other hand to produce the vacuum it is requisite that at the end of the stroke the cylinder should be brought to the ordinary temperature, say 60° F. How were these opposing requirements to be reconciled?

Watt made experiments on the heats at which water boils under different pressures, and also on the volume of steam at atmospheric pressure resulting from a given volume of water and he found it was about 1800 times as great—a fairly accurate result. He made a boiler which showed the quantity of steam used at every stroke of the model and found it to be several times the volume of the cylinder. He passed steam into a quantity of water in a glass vessel until the water became boiling hot, when he found it had gained about one-sixth in volume; in other words water converted into steam can heat about six times its own weight of water from room temperature to boiling point. He was puzzled by this result until he learnt from Professor Joseph Black of the latter's discovery in 1761 of the property of latent heat, viz. that bodies changing their physical state either absorb or give out heat at the moment of change, for example, ice melting into water or water boiling into steam absorb heat. The thermometer meanwhile shows no sensible fall in temperature. As Watt said: "I thus stumbled

3-2

upon one of the material facts by which this beautiful theory (i.e. of latent heat) is supported."

Where and how Watt, after turning over the problem of keeping the cylinder hot and yet condensing the steam hit upon the idea of the separate condenser in May 1765, is best told in the words of one who had it from Watt's own lips, albeit half a century after the event; it is a matter for regret that we have no contemporary account of it. The narrator was Robert Hart, an engineer in Glasgow, a hero-worshipper of Watt, and the occasion was when the latter was visiting the city in 1813 or 1814:

It was *in the Green of Glasgow*. I had gone to take a walk on a fine Sabbath afternoon. I had entered the Green by the gate at the foot of Charlotte Street—had passed the old washing-house. I was thinking upon the engine at the time and had gone as far as the Herd's house when *the idea came into my mind, that as steam was an elastic body it would rush into a vacuum, and if a communication was made between the cylinder and an exhausted vessel, it would rush into it, and might be there condensed without cooling the cylinder.* I then saw that I must get quit of the condensed steam and injection water, if I used a jet as in Newcomon's engine. Two ways of doing this occurred to me. First the water might be run off by a descending pipe, if an offlet could be got at the depth of 35 or 36 feet, and any air might be extracted by a small pump; the second was to make the pump large enough to extract both water and air.....*I had not walked further than the Golf-house when the whole thing was arranged in my mind.*

The experiment that followed and others that had preceded it were done in a workshop off King Street.

On being asked more precisely as to its situation Watt replied: "It was in a little court, near the end of the Beef Market, the house projects into the Court; I think a carter occupies it at present." Robert Hart and his brother John went there the following morning (we are speaking it is to be remembered of 1813 or 1814) and identified the shop: it was where Miller's Place now stands.

Since it was the Sabbath day Watt, bowing to the convention of the time, had perforce to wait until the morrow, when he set to work and knocked up an apparatus with which he proved to his own satisfaction that his idea was correct and that steam would rush into a vacuum, as he had reasoned it would. This apparatus we believe to be that preserved among the Watt models in the Science Museum, South Kensington, where it was received in 1876 along with other models, presently to be described, from James Watt & Co. of Birmingham, to whom was handed down the tradition that it was the identical one employed by Watt (see Pl. IV). It tallies very well with his description of his experiments and certainly bears every evidence of hasty improvisation, e.g. it is made of tinplate and solder while the top of the condenser F is closed by a sewing thimble taken, we imagine, from his wife's workbasket.

As the model now exists, it has been soldered together in such a manner, whether by accident or design, that it is unworkable. The diagram (Fig. 2), prepared from the model, shows how by a slight shift of the connections, the condenser can open into the cylinder and not as at present into the steam jacket. Referring to this diagram, the cylinder A has around it an annular space or steam jacket in communication with a boiler by the opening B. There is a drain hole C in the bottom corner for condensate. In the cylinder works a piston D, the rod of which has a hook at the lower end on which to hang a weight. Steam was admitted into the cylinder through the opening E (now closed) in the top cover, filled the cylinder, and blew out the air through the upper part of the condenser F through the flap or clack valve G (now missing). The condenser is supposed to be flooded with cold water, and when all the air was blown out, the cock was shut and the air pump bucket in H was drawn up smartly by hand, thus exhausting the chamber F; consequently the steam from the cylinder A rushed in, was at once condensed and caused the piston to rise, and with it the weight. This is an experimental apparatus capable of a few strokes only, because the condenser water would soon get too hot to function.

The problem now was to make the apparatus into an engine capable of repeating its motion indefinitely. Watt started on the construction of a model with a cylinder 2 in. diam. While thus engaged Robison's story is that he burst into Watt's parlour and found him with a "little tin cistern" on his lap. Robison began to talk engines, as he had done previously, but Watt cut him short by saying: "You need not fash yourself about that, man; I have now made an engine that shall not waste a particle of steam." Robison put to Watt a leading question as to the nature of his contrivance but "he

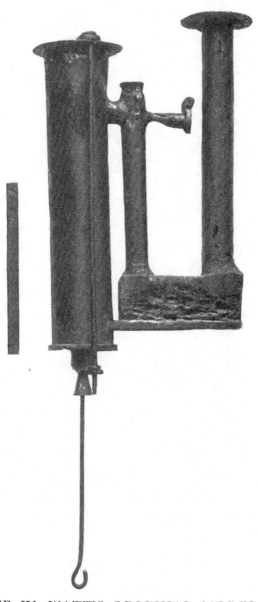

PLATE IV. WATT'S ORIGINAL MODEL OF
HIS SURFACE CONDENSER ENGINE, 1765
Courtesy of the Science Museum

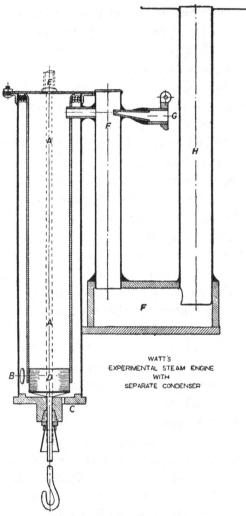

WATT'S
EXPERIMENTAL STEAM ENGINE
WITH
SEPARATE CONDENSER

Fig. 2. Diagram of Watt's surface condenser model, conjecturally restored

Courtesy of the Science Museum

answered me rather drily and vouchsafed no explanation". If an artist ever wishes to paint a genre picture of
Watt, instead of perpetuating the unfounded story of his
playing as a boy with the steam issuing from the spout
of a kettle, he might limn the young workman in his
leathern apron with the separate condenser on his lap
and Robison trying to quiz him.

If Watt lacked experience in the construction of
engines "in great", i.e. of full size, he had the advantage
of being free from preconceived ideas of what engines
should be like. In fact he had in view two engines, one
reciprocating and the other rotary, entirely different in
design from anything that had gone before. This very
fertility of mind, and his resource in expedient, may
almost be said to have delayed his progress. Of a
number of alternatives he does not seem to have had the
flair of knowing which was the most practicable, hence
he expended his energies on many avenues that led to
dead ends. In truth this is the attitude of mind of the
scientist rather than that of the craftsman. Still, unless
he had explored these avenues he could not be certain
that they led nowhere.

Watt's reciprocating engine differed from that of
Newcomen in having an inverted cylinder actuating
pumps directly; the condenser was made of narrow cells
of tinplate, closely packed, or else of small tubes, in both
cases with cooling water circulating between. His ideas
are well shown by the drawing he prepared about 1765
or 1766 for an engine that he intended to have erected at
Kinneil (see Pl. VI, p. 58).

The rotary engine, or as Watt termed it the "circular engine" or "steam wheel", which he schemed at the same time was on a different principle for with it he intended to obtain rotative motion without the intervention of reciprocating parts. We hear of it first in a letter to Roebuck (1766, Feb. 19). What this rotary engine was like is shown on a drawing, subsequently prepared for the patent specification (see Fig. 3, p. 53). The engine consisted of an annular chamber with three flap valves hinged to the inner surface so that they can turn through 90 degrees and form one abutment for the steam introduced and exhausted at the side by valve boxes, communicating with the boiler and condenser respectively through the hollow trunnion. The other abutment for the steam was to be furnished by mercury or fusible metal, the weight of which kept it constantly at the lowest point of the wheel while it revolved. Watt seems to have retained belief in his steam wheel for a long time, for he submitted a drawing of a somewhat improved design to Parliament in 1775, when evidence was given that such an engine had been constructed and tried at Soho.

It took Watt about a couple of years of thought and experiment, 1765–66, to arrive at the decision to employ the existing type of beam engine—but even so, structural changes in the engine were necessitated. To keep the cylinder hot—the *sine qua non*—he decided to substitute boiler steam for the cold atmosphere acting on the top of the cylinder. This involved closing in the top by a cover, provided with a stuffing box for the piston rod to pass through. This was a new departure, for which Watt has entire credit. Having done this he could not use water on the top of the piston for packing as previously practised and he had to devise some other means of packing. The condenser he adopted had a plain jet of water like that previously used in the cylinder of the common engine. The air pump to clear the condenser of air and water was an adaptation of existing practice. Such was the design which Watt now intended to adopt in practice.

We must bear in mind continually that Watt was a poor man and obviously could not neglect his daily bread for experimental work. Here we are happy to say that Dr Black proved a true friend, by encouraging Watt, by lending him money to carry on experiments and, what proved to be more important still, by introducing him to Dr John Roebuck of Birmingham, who has already been mentioned. Roebuck was a captain of industry who among other activities had established the leaden chamber process of making sulphuric acid commercially, and was now busily engaged in developing the coal-field that he had leased from the Duke of Hamilton, at Borrowstouness (Bo'ness) and the blast furnaces at Carron, Stirlingshire. He was experiencing trouble with water in his coal pits and was consequently interested in anything which might relieve him of this anxiety.

Roebuck was sagacious enough to see that there was something in the invention. He was not the man to let the grass grow under his feet and urged Watt on. In

August 1765 the latter gave Roebuck particulars of a trial of a model he had made. In a letter, September 19th, Watt points out the advantage of working with steam above atmospheric pressure. On October 2nd he reports the results of a trial of one of the colliery atmospheric engines—quite probably this was Watt's first introduction to one—which showed that the cylinder used about $4\frac{1}{4}$ times its volume of steam at every stroke. Watt made further experiments with a brass cylinder 2 in. diam., after which he proceeded to make an inverted engine, somewhat like Pl. VI, with a cylinder 6 in. diam.

His painful struggles with inferior workmanship, poor materials and lack of experience now continued. Piston packing was and remained for many years a difficulty. Watt tried "English pasteboard made of old ropes instead of paper and oiled". The plate condenser "came unsoldered and the drum condenser was substituted". Mercury tried as a lute "found its way into the cylinder and played the devil with the solder" as one would expect.

Roebuck never seems to have suggested that Watt should devote himself exclusively to the development of the engine; that would have meant paying for Watt's maintenance while doing so—instead of which Watt had to go on working for his living. John Craig, Watt's partner, died in 1765, thus bringing their partnership to an end and the latter probably had to refund Craig's trustees for the borrowed money. Turning round for other avenues of employment, Watt became aware of the openings for surveyors afforded by the construction of navigations or canals for the ready conveyance of goods.

In England the Duke of Bridgewater had shown the way by his canal from Worsley to Manchester, and the desire to adopt such cheapened means of carriage for goods spread rapidly through the country. Scotland, although from the configuration of the land not so suitable for canal construction as England, yet had several practicable routes and was equally desirous of canalising them. The difficulty was rather that her people were too poor to bear the capital cost. However, even to prepare schemes required the services of land surveyors, able to level, to take out quantities of earthworks, to design locks and build bridges of small span. How Watt acquired the training to take up such work we do not know. In those days simple chain surveying was taught at school and Watt may have received a smattering of this art, but levelling and the rest were professional work. All we can say is that Watt set up as a surveyor in the summer of 1766 and took an office in King Street, parallel and next to the Saltmarket westwards. We surmise that he kept on the instrument-making business and possibly the shop, because journeymen's books up to 1771 exist and no record of the disposal of either business or shop has been found.

Very soon, that is to say in October 1766, we find Robert Mackell, another surveyor, and Watt engaged in making a survey for a canal projected between the Firth of Forth at Bo'ness and the Firth of Clyde at

Dumbarton, by the Loch Lomond passage. Their report appeared in print in 1767 and in March of that year Watt travelled to London in connection with the promotion of the requisite Bill in Parliament. Here he made the acquaintance for the first time of a Committee of the House but he did not succeed in convincing them that the project was one to merit their sanction. It was perhaps this reverse which caused him to write to his wife: "I think I shall not long to have anything to do with the House of Commons again—I never saw so many wrong-headed people on all sides gathered together...I believe *the Deevil* has possession of them." Let it be said in extenuation that this was the Unreformed House of Commons.

To us the chief interest of the journey is not the promotion of the Canal Bill but the fact that he travelled, not by the usual route, but viâ Birmingham, no doubt furnished with introductions from Roebuck to the latter's old friends there, Dr William Small and Matthew Boulton. Watt's immediate purpose in visiting the town, to judge by his *Journal*, was to visit Samuel Garbett, Roebuck's partner. Watt arrived on March 16th, 1767, and not finding Garbett there "proceeded to Oxford the same night"; hence he could hardly have seen anyone. Had such a meeting taken place, it could not have been otherwise than an interesting one, if only for the reason that Small, Boulton and the great Dr Erasmus Darwin had been thinking and talking for some time past about fire-engines, the reason being that Boulton was already experiencing difficulties at Soho during the summer months owing to shortage of water to turn his water wheel, and thought that a fire-engine might be employed to pump the tail water from the wheel back into the pool. On his return journey Watt called at Lichfield and did see Dr Darwin, to whom he revealed, under promise of secrecy, his invention of the separate condenser.

Watt took occasion on this journey to visit the Bridge-water and Calder Canals (Aire and Calder Navigation) —the two show places—clearly with the view of improving his knowledge of canal engineering in which he had now launched out. It is safe to say that, had the steam-engine never been heard of further, Watt would have attained a place in the front rank of civil engineers.

In his new occupation the following winter proved to be a slack time and he turned again to the engine. Between January and May he made a large number of experiments with new designs, details of which he gives in a number of letters to the Doctor. These were so far successful that Roebuck was now obviously impressed and agreed to become partner with Watt in the project; in return for a two-thirds' interest he took over Watt's indebtedness to Black—about £1000—and undertook to pay the cost of a patent, then about £120; Watt of course was to give his time and abilities. Consequently in July 1768, Roebuck sent Watt to London to secure protection for the invention. This time he went by coach, a service that had been established in the interval since his last visit. He despatched his business and took the oath on the patent on August 9th. On the 27th he

left London, travelling by Lichfield, Newcastle-under-
Lyme and Stoke. He turned aside to Birmingham to
visit Matthew Boulton at Soho House. At once the two
men conceived a hearty liking for one another, perhaps
because of the very dissimilarity of their natures: one
the sanguine, optimistic and assured manufacturer; the
other, the cautious, pessimistic and diffident craftsman.

To realise how important this visit was, some account
of his host, however brief, must be given. Boulton, then
in his thirty-ninth year, was one of the best known men
in the Midlands. His father had been a silver stamper
and piercer at Snow Hill, and this business young
Boulton entered on leaving school. Soon he extended it
by adding the manufacture of "steel gilt and fancy
buttons, steel watch chains and sword hilts, plated
wares, ornamental works in ormolu and tortoise shell"
and more particularly steel buckles which were an invention of his own. His father had died in 1759 leaving
the business to Matthew who, by his marriage a year
later to an heiress, obtained command over a considerable fortune.

The premises at Snow Hill being too small for the increased businesses, he entered into partnership with
John Fothergill, a commercial man with a knowledge of
foreign markets. They took a lease of land that was then
a barren heath at Soho in Staffordshire, some two miles
north of the centre of the city and there erected in 1764
at a cost of about £9000 a magnificent manufactory
(see Pl. V) and a dwelling house for Boulton. Included
in the lease were the water rights of Hockley Brook,

which was dammed up to supply the manufactory with power by means of a water wheel.

It is to be observed that in the Midlands at this time it was exceptional to carry out work in a large factory. Since the rise of Birmingham at the beginning of the eighteenth century to the position of the centre of the hardware industry, it had been the home of the small workshop, and of the small master with a few journeymen working by the job or piece. The aggregation in one building of numerous classes of workmen under a single technical and business control was a new departure. It had within it, as will be realised, the germ of modern mass production.

Such was the organisation built up by Boulton and there he had brought together a force of some six hundred highly skilled craftsmen capable of making the various kinds already mentioned of wares demanded by the fashion of the day. With this organisation and this force, Boulton had set himself to realise his ambition, which was to remove the stigma of "Brummagem" (the popular pronunciation of the name of the town), applied to its products in the sense of showy but worthless, to raise the standards of taste and workmanship of his products and to make the name of Soho known all over the world. The similar aim of Josiah Wedgwood at Etruria will at once occur to the mind. No wonder the local poet felt the divine afflatus and burst into song; we spare the reader all but one couplet:

> Soho! where Genius and the Arts preside
> Europa's wonder and Britannia's pride.

PLATE V. SOHO MANUFACTORY

From the engraving in Shaw's *Staffordshire*, 1798

The praise, however, is not excessive, for Boulton's products were not merely transient but of permanent artistic value; his ormoulu and his plate are sought to-day by collectors and examples find a place in our industrial art museums.

It was to this fascinating place that Watt was introduced on a fine August morning in 1768. The arrangement of the Manufactory, the skill and inventiveness displayed there, appealed to the craftsman in him. Soho, he felt, was a place such as he had hitherto only dreamed of, where workmen capable of fashioning his engine were to be found. Boulton too was by way of being an inventor himself: the inlaid button was his; he was the first to apply a mill to turn the laps for polishing steel ornaments; a shaking box for scouring button blanks was also his.

Boulton and Small discussed the ambit of Watt's patent specification which now was due to be enrolled, and a possible partnership was mooted. It was a time of delightful communion indeed! Watt extended his stay at Soho House for a fortnight.

Arrived in Glasgow on October 11th, 1768, Watt lost no time in calling on Dr Roebuck with the news and suggested that Boulton should be admitted a partner to the extent of one-third. Roebuck promised to make Boulton an offer of some kind. Watt thereupon wrote on October 20th to Boulton as follows:

When you were so kind as to express a desire to be concerned in my fire engine, I was sorry I could not immediately make you an offer. The case is thus. By several unsuccessfull projects & expensive experiments, I had involved myself in a considerable debt before I had brought the theory of the fire engine to its present state. About three years ago a Gentleman who was concerned with me, [i.e. John Craig] dyed. As I had at that time conceived a very clear idea of my present improvements & had even made some trial of them, tho' not satisfactory as has been done since, Doctor Roebuck agreed to take my debt upon him & to lay out whatever more money was necessary either for Experiments or securing the Invention, for which cause I made over to him two thirds of the property of the Invention. The debt & expenses are now about £1,200. I have been since that time employed in constructing several working fire engines on the common principle as well as in trying experiments to verify the theory. As the doctor from his engagements at Bon-ness & other bussiness cannot pay much attention to the executive part of this, the greatest part of it must devolve on me who am from my natural inactivity, want of health & resolution, incapable of it. It gave me great joy when you seemed to think so favourably of our scheme as to wish to engage in it. I therefore made it my business as soon as I got home to wait on the doctor & propose you as one I wished he would make an offer to, which he agreed to with a great deal of pleasure & will write you in a few days, that if agreeable you may be a third part concern in paying the half of the cost & whatever you may think the risque he has run deserves, which last he leaves to yourself. If you should not chuse to engage on these terms we will make you an offer when the whole is more perfect which I hope it will soon be. The objections against the engine before the tryal made last winter were: that it might be difficult to make a piston sufficiently steam tight without water or some other fluid above it. Water cannot be used because if any of it got in, it would boil & occasion a loss of heat. Oil of some kind was the next thought of but possibly it might be destroyed in time by the heat. The next objection was that the vacuum might not be formed by exhaustion sufficiently sudden or sufficiently perfect.

If Doctor Small should chuse to be concerned with you in this, I have occasion to think it would be agreable to Doctor Roebuck and would be highly so to me. If you should not chuse to engage with this affair in its present state, at any rate you will lett this Letter remain a secret except to Doctor Small.

Boulton's reputation stood so high with all who knew him that his favourable judgment was in itself a valuable asset. Roebuck was spurred on by it and wrote to Watt that he wanted "much effectually to try the machine at large. You are letting the most active part of your life insensibly glide away. A day, a moment, ought not to be lost. And you should not suffer your thoughts to be diverted by any other object, or even improvement of this, but only the speediest and most effectual manner of executing one of a proper size according to your present ideas." Thus exhorted, Watt furnished him on November 9th with details of an engine which he proposed to erect at Kinneil; meanwhile he was also busy drafting the specification and corresponding with Small about it.

The memorable patent of Watt, describing himself as a "merchant", for "a new invented method of lessening the Consumption of Steam and Fuel in Fire Engines" was sealed on January 5th, 1769 [No. 913]. It covered England and Wales, Berwick-on-Tweed and the Plantations in America. Small and Boulton advised Watt badly over the specification and in consequence he made two serious errors of judgment, firstly in not appending drawings, although he had prepared them, and secondly in patenting a principle of action and not an application of a principle. This second error led, as we shall see, to a peck of trouble later, when the validity of the patent was assailed.

The drawing which Watt had prepared for the specification is here reproduced (see Fig. 3).

Roebuck was now keener than ever on the engine patent and so convinced of its value that he was not disposed to do more than offer Boulton a licence to make the engine in the Midland Counties only; the latter, however, was not attracted by such a limited field for exploitation; his idea was the much grander one of building for the whole world. His letter on this point (1769, Feb. 7) shows such a masterly grasp of the position and insight into the future that we must quote it *in extenso*.

...the plan [i.e. Roebuck's three-county licence plan] proposed to me is so very different from that which I had conceived at the time I talked with you upon the subject that I cannot think it is a proper one for me to meddle with as I do not intend turning engineer. I was excited by two motives to offer you my assistance which were love of you and love of a money-getting ingenious project. I presumed that your engine would require money, very accurate workmanship and extensive correspondence to make it turn out to the best advantage and that the best means of keeping up the reputation and

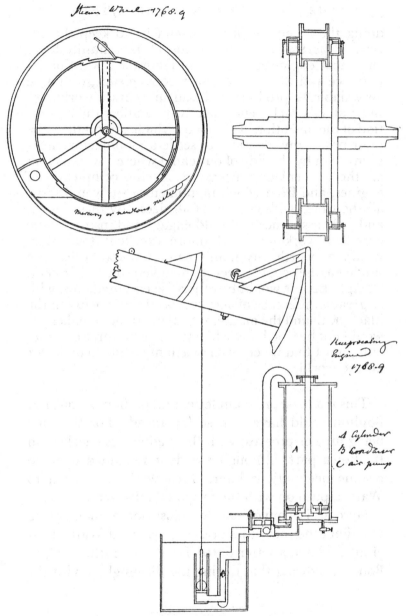

Fig. 3. Drawing prepared for the Patent Specification, 1769

Doldowlod Papers

doing the invention justice would be to keep the executive part out of the hands of the multitude of
empirical engineers, who from ignorance, want of experience and want of necessary convenience, would be
very liable to produce bad and inaccurate workmanship; all of which deficiencies would affect the reputation of the invention. To remedy which and produce the
most profit, my idea was to settle a manufactory near
to my own by the side of our canal where I would erect
all the conveniences necessary for the completion of
engines, and from which manufactory we would serve
all the world with engines of all sizes. By these means
and your assistance we could engage and instruct some
excellent workmen (with more excellent tools than
would be worth any man's while to procure for one
single engine) could execute the invention 20 per cent.
cheaper than it would be otherwise executed, and with
as great a difference of accuracy as there is between the
blacksmith and the mathematical instrument maker. It
would not be worth my while to make for three counties
only, but I find it very well worth my while to make for
all the world.

This is a most pregnant letter and perhaps no one but
Boulton could have seen so far ahead. For the time
being further progress with the engine was checked on
Boulton's part, although the door remained open to
resume negotiations later. Meanwhile it was left to
Watt to push on with the proposed full-sized engine.

Here it is convenient to pause for a moment to
mention other activities of the many-sided Watt. First
of all we have a letter dated December 12th, 1768, to
Boulton showing that during the Birmingham visit the

latter must have placed an order with Watt for a
Papin's digester and a few other trifles, evidently for
experimental work. The letter is as follows:

I have packt up for you & sent by way of Carron &
London 1 Doz German flutes at 5/-, a Copper digester
£1. 10. The digester has one side a place with 3 holes in
its lid, one for the thermometer, one for a gage pipe (the
lower end of which is to be placed in a cistern of Mercury)
& a safety valve. The steel-yard and weight for the
valve are along with the thermometer. [The latter] may
be made tight by wrapping a bit of paper about the neck
of it and thrusting tight into the hole from the inside.
I have sent in the same box some bitumastic pipes for
Doctor Small, also a piece of petuntse & a lump of im-
pure Kaolin which is excellent for stopping the pipes
with as it does not fly tho' thrust into the fire quite wett.
I have not gott the furnace made for you but shall soon.
I have almost finished a most compleat model of my
reciprocating engine; when it is tryed I shall advise the
success.

The first part of the letter gives the impression that
Watt was still keeping on his shop, though the words
"I have not gott the furnace made for you" might
suggest that the workshop had been handed over to
someone else. The mention of "petuntse", which is the
powdered crude kaolin or raw material used by the
Chinese in their manufacture of china, is evidence of the
exhaustive way in which Watt must have gone into the
manufacture of porcelain. Incidentally it should be
noted what a long way round the goods ordered had to
go to reach their destination.

It was a trait in Watt's make-up that he never had to

do with anything without envisaging means of improving it. This is well shown in the way in which he sought to enlarge the resources available for carrying on work in his new profession. One of his ideas was for a telemeter. Let him describe the idea in his own words in a letter to Small (1769, Jan. 28):

I have many things I would talk with you about much contrived & little executed. How much would health & spirits be worth to me!

I have contrived a most excellent method of measuring distances by means of a tel^e:

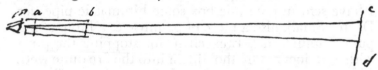

a.b. is a telescope with two paralel crossed hairs in the focus of the eye glass, c.d. is a pole divided into feet and inches. If the hairs comprehend one foot at one chain distance they will comprehend ten feet at ten chains &c. I have tryed this experimentaly & can measure half a mile to a yard; don't publish this unless it be printed already.

Our pottery is doing tolerably tho' not as I wish. I am sick of the people I have [to] do with, tho' not of the business which I expect will turn out a very good one. I have a fine scheme of doing it all by fire or water mills but not in this Country, nor with the present people. I have tryed no chemical experiments this winter.

As we shall learn later more about this way of measuring distances, we will reserve comment upon it till then (see p. 75).

Watt was, as we have seen, a pessimist, and sometimes got very down in the dumps. The following letter to

Small, April 28th, 1769, must be taken as typical of one of these periods, and we cannot resist quoting it if only for the last sentence, which is really a gem. How many inventors since Watt have re-echoed these words at one time or another!

...I find that I am not the same person I was four years ago when I invented the fire engine and foresaw even before I made a model almost every circumstance that has since occurred. I was at that time spurred on by the alluring hope of placing myself above want without being obliged to have much dealing with mankind to whom I have always been a dupe. The necessary experience in great was wanting; for acquiring it I have mett with many disappointments. I must have sunk under the burthen of them if I had not been supported by the friendship of Doctor Roebuck;...I have now brought the engine near a conclusion, yett I am not nearer that rest I wish for than I was 4 years ago; however, I am resolved to do all I can to carry on this business and if it does not thrive with me I will lay aside the burthen I cannot carry. Of all things in life there is nothing more foolish than inventing.

Yet the pessimistic mood did not last long, for he was up again working in the shop or going into the pottery works at Delftfield and in the evening meeting friends at the Anderston Club, picking up and imparting information on all sides, for Watt was insatiable in his thirst for knowledge and he went to any length to acquire it, even to learning a foreign language. This is shown in his letter to Small (1769, May 28):

I have now got a curious book being one [telling] of all the machines, furnaces, methods of working, profitts and of the mines of the Upper Hartz. [Could this have been Schlüter's *Gründlicher Unterricht von Hütte-werken*, 1738?] It is unluckily in German which I understand little of but am improving in by the help of a truly chymical Swiss dyer [his name is given later as Chaillet] who is come here to dye standing red on linnen and cotton, in which he is successful. He is according to the custom of Philosophers ennuyé to a great degree but seems to be more modest than is usual with them and, what is still more unusual, is attached only to his dyeing, tho' he has a tolerable knowledge of the rest of Chymistry. He promises to make me a coat that will not wett tho' boiled in water. This would be of great use to 100 people I see just now running by wett to the skin, no doubt cursing god in their hearts. I believe the drops are an inch in diameter.

It certainly can rain in Glasgow! Watt did get a waterproof or oilskin coat later and he needed it when he was out surveying in a damp climate such as that of Scotland.

A short interval of slack time in the autumn enabled Watt to take up the engine again. He reported the result of his experiments to Small at extraordinary length in a letter of September 20th, 1769: It is so full of interesting details that we cannot resist quoting a large part of it.

...I have no doubt but you would...be glad to know somewhat of the health of the engine as well as of the projector. Well, then, the tryal has not been decisive, but I am still allowed to flatter myself with hopes; you shall judge. The pump is $18\frac{1}{2}$ inches diameter & 25 feet high; the cylinder 18 inches wanting $\frac{1}{8}$ or $17\frac{7}{8}$; the boyler $5\frac{1}{2}$ feet at the bottom which is the widest part. The cylinder is then loaded with nearly

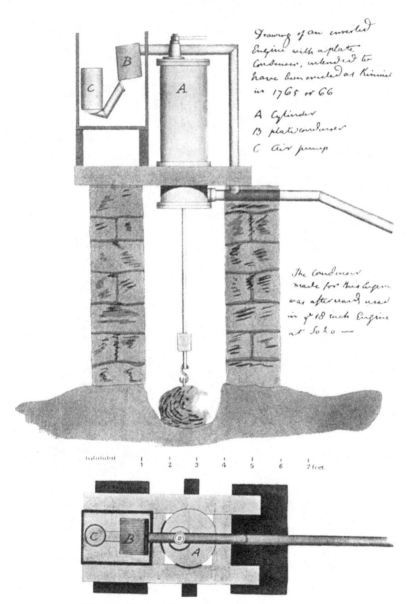

Drawing of an inverted
Engine with a plate
Condenser, intended to
have been erected at Kinneil
in 1765 or 66

A Cylinder
B plate condenser
C Air pump

The Condenser
made for this Engine
was afterwards used
in ye 18 inch Engine
at Soho —

|ₐₗₐₗₐₗₐₗ| 1 | 2 | 3 | 4 | 5 | 6 | 7 feet

PLATE VI. INVERTED ENGINE INTENDED
TO BE ERECTED AT KINNEIL, 1765

From Muirhead, *Mechanical Inventions*, III. Pt I

12 lbs on the inch, for the area is 253.8 cub. inches [this should read 250.9 sq. inches], the load 46.1 cub. feet, or at 62 lbs per [foot =] 2858.2 lbs. The adjusting and fitting all the parts together took longer time than we thought of, but after much close labour we got it brought to a tryal about a fortnight ago. After the air was pumped out the piston of the cylinder descended about 2 feet and stopped there being willing to come no further. Steam was admitted and it ascended. On a second tryal it came down only a few inches. From some circumstances I thought the bucket of the pump was in fault. The water being let off and the bucket drawn (which was not easily done), the leather was found to be what we called flyped or turned inside out thus . On examining the piston of the cylinder, the paste-board used for leather there was found torn; it was conjectured that the jack head hole might not be in the center of the cylinder. That was endeavoured to be rectified and 3 ply of pasteboard was put on the piston instead of one, a double leather was put on the bucket & the two were pinned together. We again set to work. After the air was pumped out, the piston descended briskly for a few strokes but grew gradually slower as the water rose in the pump. When the water came to the pump-head, the piston always waited some time after the valve was open to condensation before it descended. On suffering the strength of steam to increase, the piston descended more briskly but I thought hardly in proportion to the increased pressure on it. The bucket of the pump made a groaning noise by which I thought the friction in it might be more than usual, it having been rusty a little when put on. After

some strokes the piston failed and the oil came through the condenser. The piston being drawn, cork was put on in the same manner as the pasteboard, the oil pump was examined & the passage thro' which it should discharge its oil found too small; this we could not remedy at Kinneil. On putting in the cork piston, in descending it did not apply itself to the cylinder at one place of one side. On examination the cylinder was oval in that place either from some inaccuracy in the making or from some injury received in setting it up. This also we could not immediately remedy. We drew the bucket & drove nails into the leather both to smooth the barrel & presently diminish the friction. I thought the cork piston when warm & pressed by the steam might apply itself to the cylinder notwithstanding the inequality. On trial it did so but failed after a few strokes. The pump had more friction than ever because both the leather & nails were too much thickness on the bucket at once. I took out the nails and cut the leather short. I put in a piston of cork flatwise so that it was pressed against the cylinder only by its own elasticity and not by the steam. The engine wrought rather quicker than before but raised little water as most of it got down by the side of the bucket, now too slack. The piston did not fail in this experiment & if the oil pump had in any measure done its office, would have answered, but as it was a considerable quantity of steam got past it which was heard constantly rushing into the condenser. The leather of the bucket was lengthened & the piston was changed for 2 ply of pasteboard sewed together. The pump threw good water, the engine went as well as ever but always waited a little at top before it descended. The rushing of steam was heard as usual & in this, as well as in the former experiments, some oil came through with the steam & mixed with the water which it made milky &

would not easily separate but ran away with it. The oil in the cylinder & on the piston was grown thick, ropy & white & was heavier than common oil. The oil's coming through the condenser is thus accounted for. The oil pump being incapable of returning it, part of it lay at the bottom of the cylinder where any water yt was in it at first would also lye under the oil. When the valve was opened to vacuum, this $\triangledown$ [i.e. water] would boil violently & carry some of the oil along with it. In the succeeding strokes this could not happen but the oil pump always returning some oil, though not enough to drown the piston, the steam hurrying through would drive it to dust & bring it through the condenser along with it. From some experiments I made since, I hope that some means may be fallen upon to render oil on the whole less miscible with $\triangledown$. The calx of lead dissolved in it seems to answer that purpose in some measure but had the oil been returned as intended, it would have been out of harm's way. I wish however you would consider this subject & try some experiments as it [is] of great consequence that the oil should endure some time or the use [of it] must be forbidden.

The pasteboard used for the piston appeared not to be so thoroughly soaked in the lintseed oil as what I had formerly used which, with the pistons not being exactly fitted to the cylinder, & consequently leaving too great a space for the pasteboard to cover, appears to have been the cause of its destruction. I am now preparing some better & am going to try several other things.

I have hitherto spoken only of the circumstance that are against me; I must say something of what is for me.

The boyler with a small fire easily supplied more steam than we could destroy although there were many outlets for it which we took no care to stop, being employed otherwise. The boyler top and wooden cylinder

were very tight as were all our vacuum joints & valves.
The places that let out steam were at the manhole door
and at the screws that fastened the steam box to the
wooden cylinder which, had other things been right, we
should soon have corrected.

The only conclusion I can draw from this tryal is that,
supposing we cannot employ oil to keep the piston tight
& that we cannot make it better than we had it, it would
work easily with 8 lbs on the inch and would not con-
sume above $\frac{1}{2}$ the steam used by a common engine.
Even this I will not positively affirm although I think
there is reason to believe it.

Doctor Roebuck was to set off for England as today.
I have had some conversation with him about making
Mr. Boulton & you a proper offer which I expect he will
do when with you. From what I now write you, in
which I assure you I have not concealed any circumstance that makes against the engine, you will judge how
far it may be your interest to engage in it on its own
account. As to the Doctor, he has been to me a most
sincere generous friend & is a truly worthy man. As for
myself I shall say nothing but that if you three can agree
among yourselves, you may appoint me what share you
please & shall find me willing to do my best to advance
the good of the whole, or if this should not succeed to do
any other thing I can to make you all amends, only re-
serving to myself the liberty of grumbling when I am in
an ill humour.

Another letter six weeks later is indicative of in-
tensive experimental work in the meantime on the
Kinneil engine. The letter (1769, Nov. 3) reads as
follows:

I...have been impatient to hear from you for tho',
from what I said before, I would scarcely, if it was left to

myself, advise you to engage in this scheme, yett I am selfish enough to wish you would do it of your own accord. I have been so busy surveying that I have not yett got the purposed alterations compleated, tho' they are in great forwardness.

The new condenser consists of 2 sets of pipes 8 in each sett, thus, o o o o   o o o o they are each $\frac{3}{4}$ inch diameter o o o o   o o o o &   18   inch long, 16 inch of which will be evacuated each stroke of the pump. They are to be $\frac{1}{2}$ inch distant from one another in all directions. Each set is to [be] surrounded at $\frac{1}{2}$ inch distant in a box of wood thro' which cold water can be made to run at pleasure. They are joined at top by a thin cast iron box thro' which they communicate with the steam. It is made sloping at the ends that as little useless [water] as possible may be in circulation. They are joined at bottom by another cast iron box thro' which they communicate with the pumps which are of copper 5 inch diameter. Two diaphrahms of wood prevent their communicating with one another, the box being continued only for strength. It was indeed originally made with an intention to have no communication at top but the steam to enter at a, by which means the valves needed only to be water tight whereas they must be airtight above, but on con[n] the steam whenever it entered (which would be as soon as any part of the pipes was empty) would rise or attempt to rise thro' the water in the pipes & would heat it but the piston of the pump still ascending thus,—that water would descend into the box below & part might get into the pump & be there converted into steam, not being exposed to much condensation. At best if it got no farther than the box, it would remain warm until the cold of the pipes made a vacuum, when it would boil & go to steam to be condensed & it is much to be feared that this heating & cooling might prolong the time of condensation. I have some time thought that some thing of this kind happened

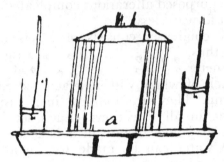

in the other condenser & concurred with other devils to plague me & indeed I was so out of humour with myself & it, I did not try all exp^{ts} I might have done to clear that up.

We have quoted these letters at great length, not only to show the difficulties with which, as we have said, Watt was faced, owing to lack of technique, indifferent materials and poor workmanship but also to show his resource in meeting difficulties, and to emphasise what is frequently overlooked, how tedious are the stages that have to be gone through before an important invention can be brought to fruition. Taken on the whole the results of the trials were inconclusive. This enumeration of the circumstances that are against him before mentioning those that are in his favour is typical of Watt, but the extremity of caution expressed in the words "Even this I will not positively affirm although I think there is reason to believe it" has not, we think, been

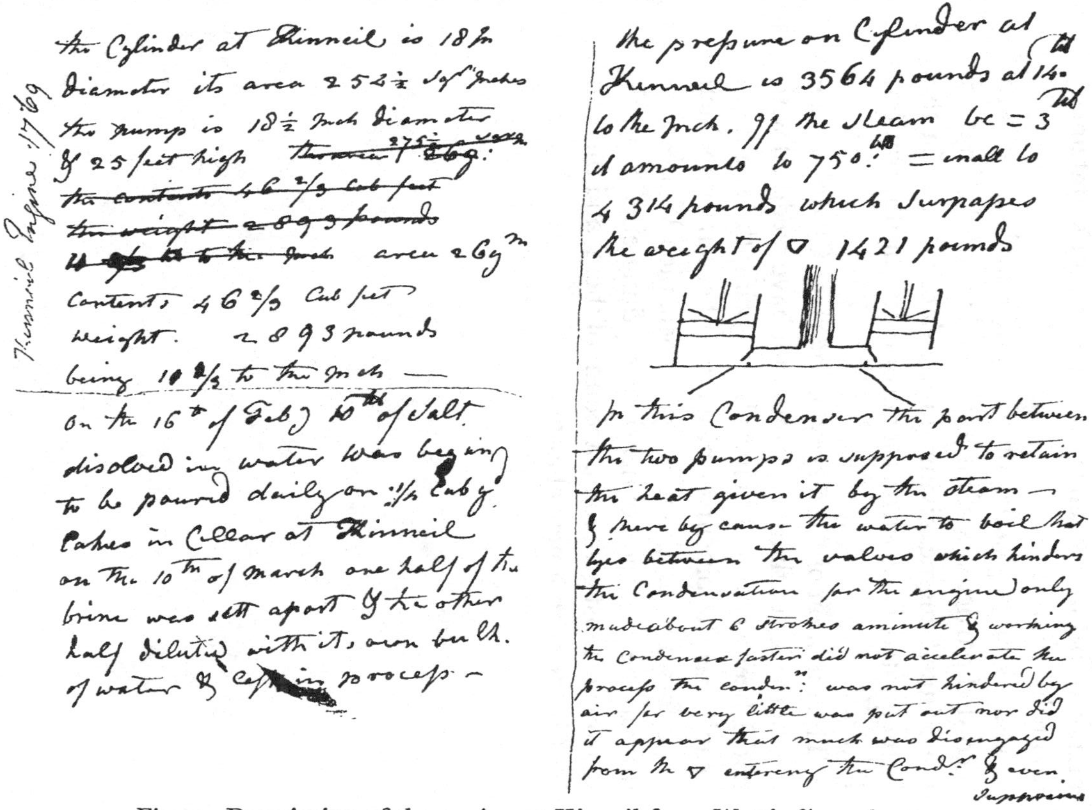
Kinneil Engine 1769

the Cylinder at Kinneil is 18 In
Diameter its area 252½ sqr Inches
the pump is 18½ Inch Diameter
& 25 feet high
~~the contents 46 2/3 Cub feet~~
~~the weight 2893 pounds~~
area 269 In
Contents 46 2/3 Cub feet
weight 2893 pounds
being 10 2/3 to the Inch —

On the 16th of Feby 10 lb of Salt
disolved in water was begun
to be poured daily on 1/2 Cub y.
Cakes in Cellar at Kinneil
on the 10th of March one half of the
brine was sett apart & the other
half diluted with its own bulk
of water & in process —

the pressure on Cylinder at
Kinneil is 3564 pounds at 14 lb
to the Inch. If the Steam be = 3 lb
it amounts to 750 lb = make to
4314 pounds which surpasses
the weight of ▽ 1421 pounds

In this Condenser the part between
the two pumps is supposed to retain
the heat given it by the steam —
& there by cause the water to boil that
lyes between the valves which hinders
the Condensation for the engine only
made about 6 strokes a minute & working
the Condenser faster did not accelerate the
process the condensn: was not hindered by
air for very little was put out nor did
it appear that much was disengaged
from the ▽ entering the Condr. & even
supposing

Fig. 4. Description of the engine at Kinneil from Watt's *Journal*, 1770
From the Muirhead Papers. Courtesy of Birmingham Reference Library

matched before or since in an inventor's statement
about his results.

Watt kept a *Journal* of his doings and in this are
recorded particulars of his experimental work. That on
the Kinneil engine, giving similar data to those in the
letter quoted, is shown on the accompanying illustration,
Fig. 4.

The mention of "a survey of the river Clyde" at the
end of the letter that he "would not have meddled with
had I been certain of being able to bring the engine to
bear" was the precursor of many other surveys which
henceforward for about five years occupied Watt's time
fully to the exclusion of nearly everything else. For this
reason we must give some account of them, and this can
best be done in a new chapter.

CHAPTER IV

PRACTICE AS A CIVIL ENGINEER,
1769–1774

Surveys and reports on canals and other works in the counties of Lanark, Argyll, Fife, Renfrew, Stirling and Inverness. Improvements in surveying instruments. Bankruptcy of Roebuck. Boulton acquires his interest in the condenser patent. Death of his wife. Removal to Birmingham.

TO an outsider it now looked as if Watt had relinquished all idea of prosecuting his engine invention and had launched definitely upon the career of a civil engineer. The survey of the Clyde estuary mentioned at the close of the last chapter was the initial step taken by the Magistrates of Glasgow with the intention of making the city into a seaport, and is of interest for that reason. Watt made a report on his survey which was printed, but it was not acted upon. The time was not yet ripe.

Another scheme upon which he was engaged in the autumn of 1769 was the survey of a canal nine miles long between Monkland and Glasgow to reduce the cost of carriage from the Lanarkshire coal-field to that city. He acted as engineer at a salary of £200 per annum till 1772 when the undertaking came to a stop for lack of funds; in fact the canal was not finished until after Watt had left Scotland.

It was while engaged on these surveys that Watt's acute mind observed defects in the surveyor's level that he used. One of these defects was in the collimation and this he animadverts upon in a letter to Small of November 3rd, 1769. These defects were of a minor kind, such as were corrected as the design of the level was improved— for the instrument was then comparatively undeveloped —and are not of sufficient importance to merit taking up the time of the reader in describing them. Another defect in the level that Watt observed was the difficulty of ensuring that it did not shift when taking a reading— say on quaking ground or in a high wind—and thereby vitiating the observation. He describes his remedy in a letter to Small (1769, Dec. 12):

If I was to follow the business of levelling I should certainly make some more alterations on levels, for, from different causes while you are looking thro' the telescope it is, possibly, altering its situation. This you cannot know till you remove your head & examine the level & if it blows or the ground is soft you can never be certain take what time you will & I assure you this circumstance kills not a little time not much to the comfort of the observer whose cold fingers & toes urge him to be gone. By the by I got a violent toothach & lost a tooth by this very thing about a fortnight ago. Now if the level would travel faster & could be seen at same time with the object it would remove this inconveniency & would render the operation more certain. I propose to take a glass tube $\frac{1}{10}$ inch dia^r. & bend it into a paralelo-gram 18 or twenty inches long & 2 inches wide & fill it with spirits to a & b then join it hermeticaly at c. Now if I see the 2 surfaces a & b in a line with any third object that object is level with my eye. Now take a four glass or day telescope whose ob[ject] glass focus $= a. b.$ cut a slitt in it in the focus of the first eye glass to admitt the leg *b* & another where it answers to admitt *a*. The leg *b* will be

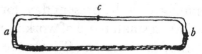

seen erect and magnified & if a 5th glass or rather a segment of a glass be introduced so as to form a telescope by which you can see the leg *a* erect & of the same size as *b* you will have the thing desired, for the motion of the

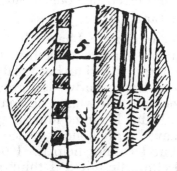

liquor in the legs *a* & *b* will be just equal to the motion of the picture on ye foc[us]: ob[ject] glass & will be magnified as much by the eye glass (you will easy conceive the adjustment necessary here). The field will have this appearance. The surfaces of the liquor have all the necessary sharpness but I have at present no time to try any further expts.

His idea was that by bringing the two bubbles of the spirit level on to the staff at the moment of observation, the surveyor could assure himself that the instrument had not shifted. Watt did not follow up his idea, but it has been brought out as new, in a neater form certainly, many times since.

In studying the life of a great inventor like Watt, one feels how valuable it would have been if he had had, as has been managed at the present day, for example, in the case of Edison, a small band of workers to gather the inventions as they are thrown off from the great man's brain and to set to and develop what he has had no time to do himself.

Yet while Watt was making these valuable suggestions for improving the level, he could write the following dismal letter to Small (1770, Jan. 31):

As to want of contrivance, I have felt severely the very thing you complain of; it was about its height when I saw you last. It is rather curious but I realy think that on the whole I grow less inventive as I grow older, but I have the consolation that what I do contrive generaly answers better. Today I entered into the 35th year of my life & I think have hardly done 35 pence worth of good in the world but I cannot help it. I do as much as indolence and ennui permit & I think myself less in-dolent than I once was.

Watt made a mistake in the date of his birthday, by the way, but that is what often happens as one grows older.

We must now revert for a moment to the engine, which had meanwhile been put up at Kinneil, in order to learn what had happened to it. Roebuck continued to urge Watt to carry on with his experiments, but the latter was unable to serve two masters, as the following letter to Small (1770, Sept. 9) clearly shows:

After so long a silence on both sides I wish to renew our correspondence by excusing myself. I think in my last I gave you an account of some unsuccessfull experiments on the engine & told you I was about making some others from which I hoped better—I also wrote you of my having undertaken a survey in Strathmore for a canal. I was obliged to go to that Survey before the exp^{ts}. were compleated. My stay in Strathmore was much longer than expec^n. I had to examine & survey a country of thirty six miles in length & to hunt about for a course for a canal through country where nature had almost done her utmost to prevent it; indifferent health, & weather viciously cold & stormy were the attendants on my survey. The winds from the snowy Grampians & snow even in the valleys a foot thick on the 10th of May, convinced me of the utility of what I was about, for nothing can be more dismall than such weather in a country which nature & art have deprived of fuel—I was in that country six weeks. On my return I was obliged to attend another disagreeable operation viz: the removing my household furniture & utensils to another house—When this was over I had just gott to Kinneil to finish my experiments on the engine when on a day's warning I was called to Glasgow to attend a survey of the river. Before that was finished, the canal I had projected last winter for which an Act had been obtained was wanted to be begun under my inspection— I had now a choice whether to go on with the experiments on the engine, the event of wch. was uncertain, or to embrace an honorable & perhaps profitable employment attended with less risque of want of success— to carry into execution a canal projected by myself with much trouble, or to leave it to some other person that might not have entered into my views & might have had an interest to expose my errors (for everybody committs them in those cases.) Many people here had conceived a much higher idea of my abilities than they merit, they had resolved to encourage a man that lived among them rather than a stranger—If I refused this offer I had little reason to expect such a concurrence of favorable circumstances soon—Besides I have a wife & children & I saw myself growing gray without having any settled way of providing for them—There were also other circumstances that moved me not less powerfully to accept the offer which I did, tho' at the same time I resolved not to drop the engine but to prosecute it the first time I could spare.

Nothing is more contrary to my disposition than bustling & bargaining with mankind, yett that is the life I now constantly lead, Use & a little exertion render it rather more tolerable than it was at first but it is still disagreeable. I am also in a constant fear that my want of experience may betray me into some scrape, or that I shall be imposed upon by the workman, both which I take all the care my nature allows to prevent. I have been tolerably lucky yett; I have cut some more than a mile of the canal besides a most confounded gash in a hill & made a bridge & some tunnels for all which I think I am within the estimate, notwithstanding the soil has been of the very hardest, being a black or red clay engrained with stones. We are out altogether £450—of which about £50 for utensils. Our canal is 4 feet water & 16 feet bottom. I have for managing the canal £200 per annum. I bestow upon it generaly about 3 or 4 days in the week during which time I am commonly very busy as I have above 150 men at work & only one overseer under me, beside the undertakers, who are mere tyros & require constant watching. The remainder of my time is taken up partly by headaches & other bad health & partly by consultation on various subjects of which I can have more than I am able to answer & people pay me pretty well; in short I want little but health & vigour to make money as fast as is fitt.

Watt's painful indecision is well shown in the above letter. He never would take chances, he always played for safety, and was quite lacking in enterprise. It looks really as if he had less faith in his engine than had his friends. Will it be believed that during the three succeeding years we have found no mention of "experiments" or indeed work of any kind being carried out on the engine? Watt was so closely occupied with his survey work, which increased rapidly, that he could not find any time to devote to the engine, and consequently it came to a dead stop.

The Strathmore canal survey that he mentions was one of those public works that were commissioned by the Trustees of the Forfeited Estates (i.e. the estates forfeited after the Stuart rebellion of 1715). The canal was to follow the valley—Strathmore—from Perth viâ Coupar Angus to Forfar, but it was not carried out; much later the present railway was constructed approximately over the same route.

By way of indication of the scale of payment for professional services, for comparison with that of the present day, we can instance the Strathmore survey; Watt only charged £80 for forty-three days' travelling and in the field; for the preparation of the report and of the map he charged a further £30 ! It is but fair to state, however, that this was much below the scale of charges of such a man as Smeaton.

An interesting incident that occurred in 1770 must be

mentioned—an invitation, received through his friend Robison who was then in Russia, to go thither in the capacity of "Master Founder of Iron Ordnance to her Imperial Majesty", the empress Catherine II, at a salary of £1000 per annum. Very wisely Watt did not entertain the offer, observing to Robison: "I find myself, however, obliged to decline acceptance of it, both from being sensible that I by no means merit your recommendation and from my advancing in the estimation of mankind here as an Engineer." This is the earliest record of his having so described himself. He received another and still more flattering invitation in 1774, but by that time he was already launched on his career at Birmingham.

Reverting to the civil engineering work we should mention that with the survey of the Clyde went plans for docks and a harbour at Port Glasgow, 1769–72. In 1771 he reported on improvements of the harbour of Ayr, which were duly carried out. In the same year he surveyed routes for canals to cut through the isthmus of Crinan—Loch Gilp to Loch Craignish, and that of Tarbet—Ardrishaig on Loch Long to Arrochar on Loch Lomond. The former was eventually carried out but not the latter.

He was next employed by the authorities of his native town, Greenock, in 1772, to make a survey for "supplying the inhabitants with fresh and wholesome water" from a neighbouring burn. An Act of Parliament was obtained in 1773 and under this Act two small reservoirs were constructed at the head of Lynedoch Street. Water was conveyed from them in wooden pipes to a cistern in the Well Park. One of these reservoirs—the Berry Park Reservoir—is still in use for the supply of unfiltered water for industrial purposes.

In 1773 Watt made a survey for a canal for coal traffic in the Mull of Kintyre from Machrihanish Bay to Campbelton, this again was not carried out. In the same year he made other surveys, viz. a canal from Paisley to Hurlet; the channel of the water of Leven; and a navigation of the rivers Forth, Gudie and Devon, surely most strenuous employment for one who did not enjoy anything like robust health!

His most important employment, also in the year 1773, was to ascertain if it was practicable to construct a canal through the remote and wild tract of country with its chain of lakes between Inverness and Fort William; in other words, from the North Sea to the western ocean. It was currently believed, as it was later in the case of the Suez Canal, that there was a difference in the sea levels at the two ends! We shall have occasion to say something further about this survey later, so that for the moment we pass on.

We have mentioned above that in 1769 Watt had the idea of measuring distances by means of a telescope although in a somewhat rough and ready way. What put the idea into his head was the difficulty he experienced in making actual measurements on the very uneven ground such as he had occasion to survey. He revolved the problem in his mind and about 1771 improved upon his previous idea by the invention of what he called a "micrometer", the usefulness of which he proved in his surveys of the proposed canals of Crinan, Tarbet, and Inverness to Fort William. This micrometer, or telemeter as it would be described nowadays, comprised a telescope with a magnification of ×8; in the focus of the eyepiece he placed one perpendicular and two horizontal hairs about 0·1 in. apart. A white disc about 8 in. diam. with a horizontal red stripe 1 in. wide was fixed on the levelling staff about a foot from the ground; a similar disc was moved up and down the staff till the horizontal hairs coincided with the red stripes on both discs. Now if the distance of the staff was say 20 chains and the distance between the centres of the discs was divided into twenty divisions, each of them represented one chain. The staff he used was 12 ft. long and with this he could measure 30 chains, the divisions being about $4\frac{1}{2}$ in. each. All he had to do was to send his assistant with the staff to the point whose distance he wished to measure and get the assistant to move the upper disc till the stripes coincided with the hairs in the eyepiece. Watt found it accurate to within 1 in 100. For ranges over 5 chains, the inaccuracy arising from equal divisions is negligible; with shorter ranges this inaccuracy is measurable, so Watt calibrated his staff by making observations at each chain length and adjusting the focus by the eyepiece to suit. We claim that Watt was the first to design and make a telemeter. Alas! *tulit alter honores*. A Mr W. Green submitted an "Instrument for taking distances at one Station"—really the same invention—to the Society of Arts and in 1778 was awarded ten guineas for it although the Society was apprised of Watt's prior invention. It has been reinvented and used many times since, in an improved form, of course, because of its value in situations such as it was designed for, i.e. where obstructions intervene. Perhaps it ought to be remarked that in the instrument known to-day as a telemeter or more usually as a rangefinder the angles subtended by the object from the ends of a short base are measured at the observer's station, a much more convenient arrangement.

Watt's "cross-hair micrometer" proved unsatisfactory at a distance greater than that at which the rod could be read by the surveyor, so that in 1772 or 1773 Watt invented another one, but we need not go into this as he did not get beyond making a model in wood.

During this intervening period, we must not imagine that the steam-engine was ever entirely absent from his thoughts. Small wrote to him, October 19th, 1771: "Nothing of late years has vexed me so much as the peculiar circumstances that have retarded your engine ...you have as much genius and as much integrity, or more, than any man I know." But what could be done? There was yet another circumstance that militated against developments of the engine and that was the commercial depression that spread over the whole country in 1772–73, due to over-speculation. Heavy failures took place in London, these affected the provinces, and in Scotland nearly every private banker failed. Watt began to reproach himself for having let

Roebuck in for the expense that had so far been incurred. He wrote to Small (1772, Nov. 7):

> Nothing gave me so much pain as the having involved Dr Roebuck so deeply in that concern....I would willingly have given up all prospect of profit to myself from it provided he could have been indemnified. He is now willing to part either with the whole or the greatest part of his property in it upon such terms as I daresay in better times you and Mr B[oulton] would have had no hesitation of accepting.

Watt goes on at great length in the same letter to discuss ways and means of effecting this and then gives a most self-revealing and self-depreciatory character of himself. This is the part in question of the letter:

> Our canal [i.e. the Monkland] has not stopt but is likely to do so from our having expended the subscriptions of £10,000 upon seven miles of the navigation and having about 2 miles yet to make. We have however made a canal of 4 ft. water for one of 3 ft. subscribed to, and have also paid most abominably for our land.
> I decline only being the manager & not being engineer. I wrote you before how grievous that first part of the business was to me and it daily becomes more so. Every thing has been turned over upon me & the necessary clerks grudged to me. I am also indolent and fearfull, terrifyed to make bargains & I hate to settle accounts. Why therefore shall I continue a slave to hatefull employment while I can otherwise, by surveys and consultations, make nearly as much money with half the labor and I realy think with double the credit; for a man is always disgraced by taking upon him[self] an employment he is unfitt for. I have no quality proper for

this em^t. but honesty which reproaches me for keeping it so long.

Remember in recommending me to business that what I can promise to perform is to make an accurate survey & a faithfull report of anything in the engineer way, to direct the course of canals, to lay out the ground & to measure the cube yards cut, or to be cut, to assist in bargaining for the price of work, to direct how it ought to be executed and to give an opinion of the execution to the Managers from time to time. But I can on no account have anything to do with workmen, cash, or workmen's accts., nor would I chuse to be bound up to one object that I could not occasionaly serve such friends as might employ me for smaller matters.

Remember also I have no great experience and am not enterprizing, seldom chusing to attempt things that are both great & new. I am not a man of regularity in Business & have bad health. Take care not to give anybody a better opinion of me than I deserve, it will hurt me in the end. We have abundance of matter to discuss tho' the damned engine sleep in quiet.

In a letter to Small a week later Watt amplifies what he had said about himself previously in these words:

I would rather face a loaded cannon than settle an account or make a bargain. In short I find myself out of my sphere when I have anything to do with mankind. It is enough for an engineer to force nature and to bear the vexation of her getting the better of him. Give me a survey to make and I think you will have credit of me. I can draw tolerably. Set me to construct a machine and I will exert myself. In whatever way you may wish to employ me, I shall endeavour to follow your advice.

This is a very just self-estimate; we can only add the

comment that Watt's was one of those temperaments which expand under the warmth of friendship but wither under the cold of hostility.

Small had enquired in a previous letter: "Have you lately invented many gimcracks", and to this Watt replied at the end of the above letter:

I have contrived a new micrometer made by drawing two converging lines upon glass I believe from trial it

will answer. I mentioned a dividing screw. It has a wheel fixed upon it with 150 teeth & only $\frac{1}{4}$ inch diameter; it is moved any portion of a turn, or number of turns, by a straight line rack, the teeth of which fit it without shake and is moved by the hand or foot. It divides distinctly an inch into 400 equal parts.

We hear nothing further about this micrometer.

Now as to the dividing screw: it may be remarked that at this time considerable efforts were being expended on dividing engines, both linear and circular. We need do no more than cite in this connection those of Jesse Ramsden and of Edward Troughton. In the Watt Workshop is preserved an experimental dividing engine (see Pl. VII (a)) that agrees fairly well with the description given above and may indeed be one and the same. It is obviously meant for dividing mathematical scales.

But we have not yet come to the end of his optical inventions. In a letter to Small (1773, Jan. 17), Watt says that he is "making a new surveying quadrant by reflection". Apparently it was an alteration without

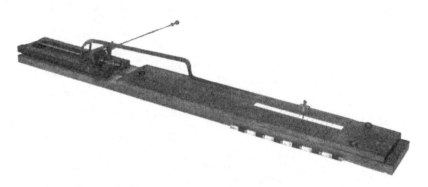

PLATE VII (*a*). SCALE-DIVIDING ENGINE
IN WATT'S WORKSHOP, *c.* 1770
Courtesy of the Science Museum

PLATE VII (*b*). WATT'S ROLLER COPYING
PRESS FOR LETTERS
Courtesy of the Science Museum

being an improvement. Small criticised it and said: "A man of your reputation must never exhibit an inferior invention", and with this dictum every one will agree.

We must now hark back to the steam-engine, for at last a step forward was about to be made with the negotiations. Roebuck's commitments at Bo'ness had brought him deeper and deeper into the mire and his embarrassment increased so greatly that in March 1773 he was no longer able to meet his obligations. From a memorandum drawn up by Watt, dated May 17th, it is clear that Roebuck had not paid more than the original thousand pounds and had not defrayed any of the payments which he had agreed to do. Consequently these had fallen upon Watt; and this is quite sufficient explanation why the latter had not been able to go on with the engine. Watt gave a discharge to Roebuck for these sums and consequently the engine at Kinneil became Watt's property. What happened is told in a letter to Small of May 20th:

Kinneil, May 20th, 1773.

As I found the engine at Kinneil perishing and as it is from circumstances highly improper that it should continue there longer, & I have no where else to put it, I have this week taken it in pieces and packt up the iron work, cylinder & pump ready to be shipt for London on its way to Birmingham as the only place where the experiments can be compleated with propriety— I suppose the whole will not weigh above four tons. I have left the whole wood-work untill we see what we are to do, conceiving it not to be worth carriage— I would not have been in such a hurry sending it off

without consulting you but it could not be delayed for
reasons I shall advise at meeting.

The position as to the ownership of the patent, however, remained unaltered—Roebuck still possessed two-
thirds of it. He owed Boulton and Fothergill some £630
and the partners would have been satisfied to have
taken over this share of the patent, on terms, but Boulton
had some delicacy in pressing Roebuck to part with it;
for one reason it might look like taking advantage of a
man when he was down and for another Boulton did not
want to foster the impression that the patent was something of great value—he determined to wait therefore
with the object of acquiring it from the trustees of Roebuck's estate. In July Watt was able to report: "None of
his creditors value the engine at a farthing." In August
Roebuck made a composition with his creditors and
eventually Boulton, by buying out Fothergill, who did
not want to have anything to do with it, got the two-
thirds of the patent into his own hands—a state of affairs
he had envisaged from the outset. His acumen in
waiting till he was able to effect this desirable consum-
mation is evident.

At this juncture, Watt was in the Highlands surveying
for the projected Inverness–Fort William canal already
mentioned. It was while so engaged that on September
26th he received the news that his wife, who was expecting her fifth child, was dangerously ill. He hurried
home by way of Tyndrum and Dumbarton only to learn
that she had died on the 24th of the month. It was a
grievous blow: "In her I lost the comfort of my life, a

dear friend and a faithful wife" and that too when more prosperous days were about to dawn. He was thus left a widower with the two surviving children, the elder only six years old.

Is it to be wondered at that he felt sick of the country, the climate and his employment, and that he should long to get out of it? The ties that bound him to Scotland were now few. His depressed state of mind is shown in the latter part of the following letter to Small (1773, Dec. 11):

I have only invented a drawing machine, the board horizontal, the index almost as long as you please & consequently the size of the picture large, a telescopic sight—no specula, the whole being performed by a most simple joint....

This damned ennui of yours is cursedly infectious. I believe like the plague it can come by post. It has seized upon me. I am not melancholy but I have lost much of my attachment to the work even to my own devices. Man's life must be spent you say either in labor or ennui, mine is spent in both. I long much to see you to hear your nonsense & to communicate my own, but so many cursed things are in the way, and I am so poor I know not when it can be.

I am heartsick of this cursed country. I am indolent to excess and what alarms me most, the longer the stupider. My memory fails me so as often totaly to forget occurences of no very ancient dates. I see myself condemned to a life of business, nothing can be more disagreeable to me, I tremble when I hear the name of a man I have any transactions to settle with.

The engineering business is not a vigorous plant here. We are in general very poorly paid, this last year my whole gains do not exceed £200 though some people have paid me very gently. There are also many disagreeable circumstances I cannot write, in short I must as far as I can see change my abode. There are two things [which] occur to me, either to try England or endeavour to get some lucrative place abroad, but I doubt my interest for the latter. What I am fittest for is a surveying engineer, is there any business in that way?

And yet in this moment of depression he could not resist inventing a drawing machine! It did not come to anything, however, for in a subsequent letter he remarks drily: "It has only one fault, which is that it will not do, because it describes conic sections instead of right lines."

He had on hand still "a survey of some improvements of the upper Forth", a survey that has been mentioned already. As to this he says: "I had many cold fingers and feet and have much boiled my brains writing a report to Lord Cathcart's genius." Then he had to finish his report on the Inverness canal. Yet will it be believed, this indefatigable man found time to make a few experiments on the heat of water at different pressures?

By April Watt had finished his report on the Inverness canal and sent it in. Generally speaking the report established the possibility of carrying out the canal, since comparatively little earthwork and few locks were entailed, and the financial aspect of it was not too formidable; however, the money was not forthcoming and consequently no action was taken.

Towards the end of 1784 the question of constructing the canal was revived and Watt was approached by his father-in-law, James MacGregor, to undertake the position of engineer. The decision was never in doubt, because Watt could not possibly have undertaken the job consistent with attention to the engine business in hand. The proposal came to nothing for the time being; but happily it was not allowed to drop. The canal was carried out in Watt's lifetime, i.e. in 1802, and is the well-known Caledonian Canal. Thomas Telford was the engineer, and in carrying it out he bore testimony to the accuracy of Watt's survey and of his description of the route.

The Inverness canal business being off his mind, there was nothing further to keep Watt from leaving Scotland and in April 1774 he was able to write: "I begin to see daylight through the affairs that have detained me so long, and think of setting out for you in a fortnight at furthest." This he did on May 17th, taking with him his family and their belongings; and arrived in Birmingham on May 31st. Boulton found accommodation for the family in his old home in Newhall Walk. Watt, now in his thirty-ninth year, entered upon the most important and brilliant part of his career.

CHAPTER V

PARTNERSHIP WITH MATTHEW
BOULTON. PERIOD OF STRUGGLE,
1774–1781

Extension of condenser patent. Partnership with Boulton. Second
marriage. Entry of pumping engine into Cornwall. William Murdock.
Letter-copying. Lunar Society. Threat to patent.

IT is not generally realised how ripe this country was
for the reception of the new steam-engine. The demand for power for mine drainage and for water
supply to towns and canals was already widespread and
although met to a large extent by the common engine
was restricted by cost of the latter in fuel. There was too
already a potential field for an engine to turn millwork;
the small water powers in which the countryside is so
rich were rapidly being exploited for the industries
already in being, or springing up on all sides. The demand for larger powers than these, i.e. such as could be
located in any desired spot, for textile manufacture or
for ironmaking, for instance, remained very largely unsatisfied.

Obviously the first thing Watt had to do now that he
had taken the plunge of giving up his profession of
surveyor, was to perfect his engine; he set to work on this
task at once and thus happily for the first time he was
able to give the matter his undivided attention. The
engine from Kinneil with its block tin cylinder was re-

PLATE VIII. MATTHEW BOULTON, *aet.* 73

From the oil painting by Sir William Beechey in the possession of the
Boulton family

erected at Soho Manufactory, where a space between the two wings of the building was allotted to him by Boulton. When the experiments were resumed the old troubles reasserted themselves. Watt's log book: "Experiments on ye first engine at Soho", noting down the results, has fortunately been preserved; we have records of the number of strokes made by the engine, the weight of coal burnt and the weight of water evaporated by the boiler. The piston and its packing was, as previously, the great obstacle: pistons of wood and of metals, packed with woollen cloth, with felt, with strawboard or even with paper pulp or horse muck and held down by weights were tried; the lubricant was vegetable oil with the addition perhaps of blacklead; these materials singly or in combinations were tried but to little effect. Eventually the tin cylinder collapsed and an iron one was sent for from John Wilkinson of Bersham. He was a well-known ironmaster, a friend of Boulton, and he had patented the previous year an improved boring mill whereby he could produce cylinders not only circular in section—that had been done previously by Smeaton —but truly cylindrical in the whole length. It was important, nay essential, that the cylinders of the new engines should possess these properties if they were to work successfully. We have some idea of what the engine looked like at this stage for we have a drawing (see Fig. 5) that was prepared by Watt for submission to Parliament in the following year, and it can be said almost with certainty that this drawing was taken from the experimental engine.

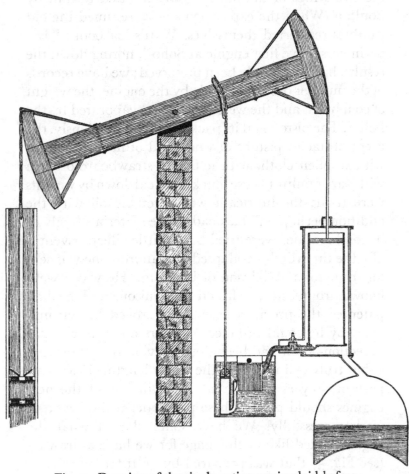

Fig. 5. Drawing of the single-acting engine laid before
Parliament, 1775

Doldowlod Papers

Eventually his efforts were crowned with success and he was able to tell his father: "The first engine I have invented is now going and answers much better than any other that has yet been made."

To make the engine work was not enough, however, there was the formidable obstacle from Boulton's point of view that the patent had only eight years longer to run and he saw no reasonable prospect of recouping himself for the necessary expenditure on it before the patent would expire. Obviously it would not be worth his while to go on with the patent unless he could see his way clearly. There were two alternatives, either to seek an extension of the patent or to obtain a new one; the former course was decided upon and the petition was presented on February 23rd, 1775; it was this business that took Watt to town. The petition met with "violent opposition from many of the most powerful people in the house". For the information of Members, Watt drew up a memorandum describing his improvements and pointing out that no monopoly was desired—only a reasonable hope of reward was expected. Boulton also came to town to help in lobbying for the Bill. To be brief, it passed all its stages and received the Royal Assent, May 22nd, 1775 (15 Geo. III, cap. LXI, pp. 1587–94). The Act extended the patent for twenty-five years and—a matter of some importance—extended it to cover Scotland.

Watt and Boulton entered into the famous partnership on the ensuing June 1st, such partnership to endure for the same length of time as the Act. Watt assigned to

Boulton two-thirds of the property in the extended patent; the latter agreed to defray all past expenses, to pay the cost of future experiments and keep the books; on the other part Watt obliged himself "to make drawings, give directions and make surveys", for which his remuneration was to be at the rate of £300 a year.

Boulton, as was his nature, threw himself with enthusiasm into the new business. He had meanwhile been raising the expectations of possible clients by telling them what the engine could do in the way of saving fuel as compared with the common one. In effect he hustled Watt to get out the necessary drawings for a large engine that could be of use for demonstration purposes. Consequently Watt started on drawings of two engines—one a pumping engine 50 in. cylinder diam. for Bloomfield Colliery, Tipton, Staffordshire, for Messrs Bentley & Co. and the other a blowing engine 38 in. cylinder diam. for Broseley Ironworks, Shropshire, for John Wilkinson. It was distinctly risky to step up from 18 in. to 50 in. diam. and there can be no doubt that it was against Watt's better judgment. He would have preferred to have made haste more slowly, but Boulton was impetuous and could not brook delay. Fortunately both engines turned out successes. The one at Bloomfield was set to work on March 8th, 1776, with much *éclat*. A long account appeared in the local paper, *Aris's Birmingham Gazette*; this is the first mention in the press of the Watt engine and indeed the first public mention of it in print of any kind. John Wilkinson's blowing engine was started about the same time but without a similar fanfare of trumpets. Judged from a mechanical point of view, these engines were ill-constructed and required a great deal of keeping in repair; nevertheless they served the purposes of gaining experience and resolving the doubts of the Thomases who were then, as they are to-day, ever with us. It is only necessary for a man to shake his head, say little or nothing, or damn with faint praise, and he achieves a reputation for sound judgment while all the time he is the fifth wheel of the coach.

It will be realised that these engines were put together on the spot, just as the common ones were; the drawings and instructions, together with a few special parts like the nozzles or valves, alone came from Soho. In spite of what Boulton had said in 1769 about mass production of engines (see p. 52), it was many years later before his ideas were realised, indeed not till 1795 when Soho Foundry was built.

Watt, while in London on the business of getting the extension to his patent, received from Boulton the sad news of the death on February 25th of Dr Small, the friend whose understanding and advocacy of the condenser had been so encouraging to Watt, and who had been in no small measure the means of bringing the partnership into being.

While in town Watt took the opportunity, at Boulton's suggestion, of measuring the performances of the common engines at the waterworks there: New River Head, York Buildings and Chelsea. Watt had made a practice of engine testing, as his note book shows, as early as 1764. Others like Smeaton had on occasion done the same, but Watt made it an everyday practice, so that it is not too much to claim for him that he is the father of systematic engine testing.

In June 1776 Watt revisited Glasgow, for he was about to take the important step of marrying again. The lady was Ann, daughter of James McGrigor or Mac-Gregor, a dyer there. Again, we know nothing of the circumstances of the engagement beyond Watt's statement: "I consider it as one of the wisest of my actions." The fact is we never really get at Watt's emotions, for he did not wear his heart on his sleeve. His prospective father-in-law, cautious man, wanted to see the deed of partnership with Boulton and the astonishing fact emerges that such a document had not as yet, a year after the partnership had been begun, been executed. This put Watt into a sad predicament and he wrote to Boulton (1776, July 3):

I have however the pleasure amidst all this to have obtained both the young lady & old gentleman's consent to take her along with me, but [I] shall be obliged to wait a little until she can be ready. This gives me pain as I am sensible I must be wanted with you, but I hope you will excuse it when you consider my situation & that it will save another journey which must otherwise have taken place soon. The only disagreeable part of the business that remains to be done is the settlement & I find that the old gentleman wishes to see the contract of Partnership between you and I, and as that has never been formally executed, I must beg the favour of you to get a legal contract written & signed by yourself, sent to me by return of post or as soon as may be. Lest he should have called my prudence in question I have been obliged to allow him to suppose such a deed did exist but was single, so what you send must pass for a duplicate and another may be actually written which I hope you will not doubt of my readiness to execute as soon as I return and in fact this deed should have been executed long ago, in common prudence upon both sides, particularly upon yours who have no legal assignations to the Act of Parliament. I therefore hope that you will excuse the old gentleman's caution. If you do not chuse to send the deed itself, you may have a scroll of it made out without date which you may send as the copy the deed was drawn from.

Boulton was equal to the occasion and replied with an ostensible letter containing—what we can only call a white lie—the statement that his lawyer Dadley had gone to London and that the deed could not be produced: instead he gave a memorandum or résumé of its contents. Had Mr MacGregor known the facts of the case, he might have been unwilling to have entrusted his daughter to the unbusinesslike Watt. We do not know positively that the deed was ever executed since it has never been found, but the incident is evidence of the absolute confidence that the partners had in one another.

Whether it was because of his marriage or the fact that he was now employed so congenially, Watt's health improved, as shown by the following paragraph in the letter just quoted:

Now for something comfortable. I have had better health since I left you than has been my lot for years and my spirits have borne me through my vexations most

wonderfully. I have lost all dread of any future connections with Monsr. la Verol [i.e. small-pox], and if I
carry my point in this matter I hope to be very much
more useful to you than has hitherto been in my power,
the spur will be greater.

It need hardly be remarked that small-pox was then a
great scourge and one could scarcely meet a person who
was not pock-marked.

On arrival in Birmingham, Watt took his wife to
Newhall Walk, where he had resided since he came to
Birmingham. Whether she was unfavourably impressed
by her surroundings, or whether Watt wanted to be
nearer Soho, probably the latter, at any rate in March
following they moved to Regent's Place, Harper's Hill,
a substantial house as will be judged from our illustration (Tailpiece, p. 122), about a quarter of an hour's
walk from the Manufactory.

Regent's Place, not the Manufactory, was Watt's
headquarters; there he did his correspondence, drawings and calculations; sometimes for days together he
would not visit Soho. It was at Regent's Place that his
assistants worked, indeed it was not till 1790 when Watt
was about to remove to his new home, Heathfield, about
which more anon, that the drawing office was transferred to the Manufactory where, indeed, it remained
till long after Soho Foundry was in full working
order.

There is no hint that Watt ever had a private workshop at Regent's Place, or that he indulged in any experimental work there, with the possible exception of

chemistry; indeed during these strenuous years he could
hardly have found time for any hobbies. He still kept
beside him, perhaps in an attic, the tools and materials
that he had brought with him from Glasgow.

The fame of the new engines spread abroad and
Bloomfield engine was followed quickly by others. That
for a distiller at Stratford-le-Bow was the first incursion
into the London area. It was this engine that Smeaton,
then a doubting Thomas, inspected in 1777 with
disastrous results as revealed in a letter to Boulton (1777,
April 20):

Smeaton said it was a pretty Engine but it appeared to
him to be too complex, but that might in some degree be
owing to his not clearly understanding all ye parts. He
gave the Engineer mony to drink & the consequences
of that was that ye next day the Engine was almost broke
to pieces. Wilbey was very angry, turned away the
Engineer & told Hadley the least amends he could make
was to put it into order again, wch he did do but was
obliged to put in one new valve.

Smeaton was perhaps hardly disinterested in his
opinion, because the improvement of the design and
construction of the common engine had been one of his
employments and he had in consequence an affection
for that engine. However, later he became so convinced
of the superiority of the Watt engine that he substituted
it for a common engine that he had been commissioned
to erect.

Bow was followed by an engine at Bedworth Colliery
near Coventry. A humorous letter written from there

by Watt to Boulton (1777, Feb. 27) just before the engine was set to work contains the following:

It is now clearly determined that Behemoth neither does nor shall stir his tail in the month of Feby 1777.— Some Engineers talk of putting in steam to try the joints tomorrow. If they do they must be busy. If they wait till the Lord's day & have his assistance, I don't know what they may do, but you need not think of coming, bringing nor sending any body here untill I bid you; otherwise they will only see a standing engine.

Apparently the partners intended to make the starting of the engine a show occasion, for below in the same letter Watt says:

It [is] now possible for you to bring the Ladies with you as the ground has got dry. If it keep so, I shall try to gett an 18-oared barge laid at Longford against the day of the firey trial.

Bedworth was started actually on March 10th; it did not turn out a particularly good engine, in fact the partners had considerable difficulty in getting the owners of the colliery to pay the premium and eventually the matter was submitted to arbitration. After the engine had been overhauled completely, a test was made in the presence of the arbitrator, and as a result he made an award of £217 per annum in favour of the partners. This was the occasion for a jubilant outburst on Watt's part—the only time we remember when he really did let himself go; having done so he seems to have repented of it immediately for as a postscript he wrote "Please burn this nonsense". The letter (1779, June 31) was not burnt, and here it is:

Hallelujah! Hallelujee!
We have concluded with Hawkesbury—£217—pr
annum—from Lady Day last, £275.5 for time past,
£117 our account.
We make them a present of £100 guineas.
Peace and good fellowship on earth—
Perrins and Evans to be dismissed—
3 more Engines wanted in Cornwall—
Dudley repentant and amendant—

Yours rejoicing

JAMES WATT.

Bedworth engine was followed by the one for Torry-
burn in Fifeshire, the order for which Watt secured it may
be observed in 1776 when he was in Scotland, on the
occasion of his second marriage. Fortunately we are
able to give a drawing of this engine (see Fig. 6), and
we do so because although not erected till 1778, it was
designed before those made for Cornwall. The title
"General Section of Engine" suggests that Watt considered it representative at the time. It will be observed
that the cylinder is steam-jacketted and below it is a
small furnace which was provided in the event, appre-
hended by Watt, that the boiler steam would not be
sufficient owing to condensation to maintain the bottom
of the cylinder as hot as it should be, but the device was
not needed. The condenser is a simple pipe kept cool by
a jacket supplied with cold water; the injection is at the
lower end. The air pump is a simple bucket. The beam is
trussed with a king-post.

The field for the engine was now widening rapidly.
Watt in a long letter from Birmingham to Boulton (1777,

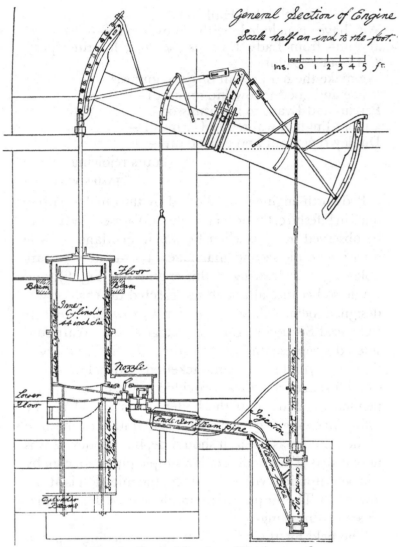

General Section of Engine
Scale half an inch to the foot.

Ins. 0 1 2 3 4 5 ft.

Floor
Beam
Inner Cylinder 44 inch dia
Nozzle
Lower Floor
Condenser Steam pipe
Ejection
Cylinder Beam

Fig. 6. General Section of Engine, 1776
Boulton and Watt Collection. Courtesy of Birmingham Reference
Library

May 2) mentions having been with Wilkinson to his ironworks at Bilston where a blowing engine was wanted. The latter, in conversation, disclosed what he had learnt was going on in France. It appeared that J. C. Périer (1742–1818), a smith by trade but a clever mechanician and an energetic man, had obtained from the French king Louis XVI an *arrêt de conseil* or decree "empowering him to raise water from the Seine to supply Paris & erecting a company". Périer's original intention had been to have a common engine from Wilkinson but after coming to England and visiting Broseley, where he learnt about Watt's engine, he realised its advantages and decided that he would prefer one of the improved kind. He tried to induce Wilkinson to make one for him, unprincipled man, insinuating that he would be out of Boulton and Watt's jurisdiction. It was, however, Wilkinson's interest to keep on good terms with one of his best customers and besides that he would have laid himself open to an action for infringement if he had acceded to Périer's request, and Wilkinson very sensibly declined the job.

The incident showed the partners that there was danger ahead and convinced them that it would be desirable to obtain a patent or a privilege in France. This they succeeded in doing through the instrumentality of the Comte d'Heronville, who had approached the firm for an engine to drain some low-lying land near Dunkirk. The firm consented to supply the engine on the condition that the Count would use his influence to obtain a privilege for France. He succeeded in his

efforts and on April 14th, 1778, the Minister issued an
arrêt de conseil for fifteen years on condition that a trial of
a Watt engine, to be made either at Dunkirk or Paris,
should establish its superiority over the common engine.

Nothing came of the Dunkirk drainage scheme, but
shortly after the decree was obtained, a certain M. Jary
who had a concession for some coal mines near Nantes,
Loire Inférieure, approached the firm for an engine for
his colliery. He obtained for them a fresh decree per-
mitting them to carry out the conditional trial there instead of at Dunkirk or Paris. Consequently, Jary was
furnished with the usual drawings, instructions and
engine parts; permission for the export of the latter was
obtained with difficulty as Great Britain was then at war
with France. We regret to have to state that, perhaps
relying upon this fact, Jary bilked the firm and never
paid them a penny piece—a deplorable example of
commercial morality. However, Jary's engine has the
distinction of being the first Watt engine in France and
indeed on the Continent of Europe.

In January of the following year Périer paid a second
visit to England, and willy-nilly had to go to Soho, where
he concluded a bargain for an engine very favourable to
himself, for the reason that the *arrêt* did not have the
effect of law while its conditions remained unfulfilled.
On his return to Paris, Périer erected two engines at
Chaillot, on the Seine outside the boundary of Paris, and
there they remained for many a long day, one of the
wonders of the city.

The partners did not need to spend time looking for

work because orders began pouring in. An extensive
field for the application of the new engine existed already in the mines of Cornwall, where pumping, owing
to the distance from which fuel had to be brought, was
the most considerable item of cost. The mine adventurers
had taken keen interest in the engine from the first, a
deputation had been to see Bloomfield engine and negotiations had been going on for some time, but it was not
till 1777 that the firm received the first order from the
duchy—an order for an engine for Ting-tang near St.
Day—quickly followed by one for Wheal Busy near
Chacewater. Wheal is the anglicised form of Huel,
meaning simply "mine".

Watt, accompanied by his wife, set off for Cornwall
in order to be on the spot. The journey was typical of
eighteenth-century travel and for that reason is worth
detailing, particularly as Watt does this in his letter to
Boulton (1777, Aug. 9):

I send Journal of my travels. Tuesday, 5 o'clock left
Birmingham; by 9 o'clock at night arrived at Bristol very
much in want of sleep having got little the night before—
Found all the places in coaches & diligences taken to
Exeter for next day & day after—so took chaise at
6 o'clock in the morning & kept company with the
diligence to Bridgewater, where places becoming empty,
we went in Dilly to Exeter. Found the town full of people
and all the stages to Plymouth full, so went on in the
Chaise way.... Thursday, a few hours at Exeter & a few
a Plymouth. Friday at Plymouth Dock, [since renamed
Devonport]—crossed over to Saltash (which I advise
you not to do), lay at Lostwithiel; Saturday Truro.

Altogether Watt and his wife were four days *en route*. His first impression of the Cornish folk, their ways and works, was anything but favourable. "The people here ...have the most ungracious manners of any people I was ever ammong".... "In general the Engines here are clumsy & nasty, the houses crackt & everything dripping with water from their house cisterns."

Watt found that some progress had been made with the new engines: "Wheal Bussy is in considerable forwardness", in fact the engine was the first to get to work; this it did in September while Ting-tang did not "fork" till July of the succeeding year.

Following upon the engines just named, two much larger engines with 63 in. diam. cylinders were ordered for Wheal Union near Marazion and Chacewater respectively. Wheal Union, as the name suggests, comprised a group of mines known as Owen Vain and Tregurtha Downs, and it was at the latter mine that the engine was installed. John Budge, or "Old Bouge" as he was familiarly known locally, was the engineer of Wheal Union, and although he was one of those who had visited Birmingham and had had the working of the engine explained to him, he appears to have failed to understand it; hence he declined to superintend Tregurtha Downs. Consequently the erection was entrusted to the firm's own erector, one Dudley, who we regret to state was a broken reed. Watt's account of the starting of this engine, given in a letter to Boulton (1778, Sept. 6), is typical of the difficulties that were encountered with these early engines and the physical exertions required to overcome them: this is the letter:

On Friday I went to W¹ Union, where I found them in ye dumps. On Thursday they had attempted to sett ye Engine to work before they had got it ready. I had told them that they had not cold ▽ for above 7 or 8 strokes pr minute. However as there was a great number of spectators, Dudley thought he would show them some what and accordingly sett off at the rate of 24 sts pr minute, he soon got all his water boiling hot and then they seemed to be at a loss why the Engine would not go, but as they found by experiment that cold ▽ improved it they sett about putting down a cold ▽ pump to addit wch was got finished by Saturday morning. D. complained that ye Barometer had choaked soon after they began but I found it was fairly soldered up & had never gone I also found many air leaks wch convinced me he had never sought after them, he had neglected to put oakum about the necks of ye regular spindles, and had used no means to prevent over opening of exhaustion regulator by wch they had had such rapps as to break one of ye spring beams. Off all these things he had been fully warned Monday & before—they sett to work at all, the fire would not burn, wh^ch was caused by a parcel of Bricks left in ye flues near ye damper I staid to 9 o'clock on Friday night and saw most things cured, I was at home again by 7 o'clock on Saturday morning and got the Engine to work by 10 o'clock quietly & peaceably 18 strokes pr minute, 5½ feet long only, being circumscribed in their working barrell, at present. It went on thus till 12 oclock that I went to Breakfast, by which time had sunk ye water in shaft 5 feet—about same time they bent the tail of ye detent of ye scoggan and I caused it to be altered to come over ye exhaustion shaft instead of under it since wh^ch has never catched they should all be made so during this interval the engine men were changed and a pair of new ones took place The

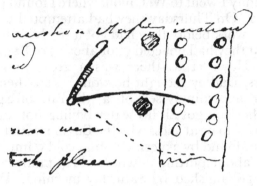

Engine was again sett to work & for perhaps half an hour went as before then ye steam failed and every endeavour to mend it was in vain though the engine was wrought by hand only 10 strokes pr minute yett ye steam was always so weak that the air went in at every crevice of ye Boiler and was voided by ye air pump As I do not thoroughly understand Cornish firing I could not say what was to be done to the fire nor whether ye fire or ye Engine was in fault, for no stranger can distinguish between a bad fire and a good one here & if they once gett bad it requires half a day to mend it I staid till near 4 oclock and then I thought it rather better but not much and as I saw nothing I could do, came home tired mortified & vexed. I shall go again tomorrow by which time expect ye first lift 10 fath^m to be forked. I charged Dudley to attend punctually and to sett about putting on the steam case tomorrow—I would not have suffered it to have gone to work without it if adventurers had not been falling off every day, there is now 1/10 of ye concern thrown up, and when an engine is so very slightly loaded I conceive heating loss to be of small moment witness commencements of Bloomfield, all ye paste-board Joints in ye Cast Iron were tight but the lead ones wept a little—I forgot to mention at first setting out yt somebody had knocked out a piece of wood which propt open ye Mediator valve yett the engine went above 20 strokes and gott ye vacuum to 28 when it took 2 men with a lever to relieve the mediator.

The engineers in charge of the common engines, such as Richard Trevithick, Senior, were suspicious of the Watt engine and resented its intrusion into their terri-tory; others like "Old Bouge" actually spread false re-ports about them, but Jonathan Hornblower, the fore-most engineer in Cornwall, was the most amenable to reason although Watt called him "an unbelieving Thomas". However, like Thomas of old, all of them found that seeing was believing and became converts to the new creed, so much so that a period of intense activity in engine construction set in and by 1780 out of forty engines built by the firm twenty had been erected in Cornwall alone and not more than one or two com-mon engines remained in the whole county.

It will be interesting to state the terms of remunera-tion that Boulton and Watt charged for their engines—terms of a novel kind that were devised by Watt himself. Writing to Jonathan Hornblower (1776, Oct. 17) Watt says:

Our profitts arise not from making the engines but from a certain proportion of the savings in fuel which we make over any common engine that raises the same quantity of water to the same height. The proportion of

savings we ask is one third part to be paid to us annually
for twenty five years, or if our employers chuse it, they
may purchase up our part at ten years price in ready
money.

Obviously where a common engine was to be replaced
by a Watt engine the fixing of the amount to be paid
could be and was determined by actual measurements,
usually made by Watt himself. Such cases were very
few in number, because ordinarily the new engine was
required to pump more water or from a greater depth
than the old one; or perhaps it was wanted for an entirely
new mine. In such cases fixing the premium had to be
done by calculation.

The first engines in Cornwall were erected without
agreements, obviously so, since a standard of comparison did not yet exist, but by October 1778 a small committee had been appointed to make the necessary test.
The common engines at Poldice Mine were accepted as
a fair average and on test the Committee found that their
"duty", i.e. the number of pounds of water raised one
foot high per bushel (usually reckoned as 94 lb.) of coal,
was upwards of seven million. It is unnecessary to
follow Watt through the intricacies of his calculations,
but his final conclusion was that the best performance of
the common engine was when the load on the piston
was 7 lb. per sq. in. while for his own engine the best
load was $10\frac{1}{2}$ lb.; this meant of course that of two
engines to do the same work, the cylinder of the Watt
engine must be smaller than that of the common
engine; for example, Ting-tang 52 in. diam. Watt

engine was equal to the common 63¾ in. cylinder whose place it took.

On the basis of the figures calculated, Watt drew up a table of sizes of his own engine equal to hypothetical sizes of the old kind. In agreements drawn up subsequently, the appropriate figure from the table was inserted as the basis on which payments were to be made. It was a fair arrangement, but it had a drawback in that the Cornish adventurers did not readily grasp the method of obtaining the figure and as a result were somewhat suspicious of it.

Having arrived at the factor whereby to calculate the savings, it became necessary to measure both the quantity of water raised and the weight of coal consumed. The latter was readily done; it was already the practice on the Cornish mines to measure the coal consumed by the engines because a drawback of the duty on exported coal was allowed by Act of Parliament, 1751, on all coal so consumed.

Measurement of the water pumped was a little more difficult, but as the diameter of the pump barrel and the length of the stroke were known, it was easy to calculate a figure which was the number of pounds weight of water delivered at each stroke. It remained then only to count the number of strokes in a given time in order to know the total weight of water lifted, that is on the assumption that all the strokes were of equal length—an assumption not always borne out by facts, so that the adventurers still had a grievance to nurse.

Obviously no one could be expected to stand by the

engine and count the strokes, so Watt designed a mechanical counter and fixed it on the engine beam. The counter resembled a clock in having a pendulum but instead of this being actuated by a spring, it was wagged to and fro by the see-saw action of the beam to which it was fixed. The counter was kept locked of course, so that it could not be tampered with.

Watt did not, as many people suppose, invent the counter. It appears that Boulton got the idea in 1777 from a pedometer that the firm of Wyke and Green of Liverpool were making. In September of that year the firm supplied " 1 long frame & wheel for counting strokes of fire engines &c. 1*l.* 5*s.* 0*d.*" Boulton made some of these counters in his works at Soho and a man named Halliwell of Birmingham made a few also. There were minor defects in the apparatus and these Watt obviated as shown in his letter (1779, Oct. 28), in which it is a delight to note how the craftsman's knowledge peeps out:

Counters will be wanted but care should be taken that the fingers be fast on the axis. Wyke wrote some time ago that they had several ready. Their counters are better than the Birmingham ones as being stronger and not so liable to break. These springs the Birmingham man has put to the crown wheel cause the counter to stop. Wyke should be wrote to to fix his needles or fingers and to harden his pallets & crown wheel pins & pinions. There must be two counters on each beam to prevent accidents.

The counter was supplied regularly after this at a cost of two guineas each. One of them is preserved in the Science Museum, South Kensington.

PLATE IX (*a*). WATT'S RESIDENCE AT PLAIN-
AN-GUARY, REDRUTH, 1778. Present day
Courtesy of W. A. Michell, Esq.

PLATE IX (*b*). COSGARNE HOUSE. RESIDENCE
OF BOULTON AND WATT IN CORNWALL, 1781
Present day
Courtesy of W. A. Michell, Esq.

The adventurers generally did not like the method of assessing the premium, so that reluctantly Watt acceded to the demand for the more readily intelligible method of a fixed annual premium for each engine according to its size. The pain which it gave Watt to abandon his carefully constructed table can be imagined; it was in truth too scientific for a workaday world.

On the whole Watt's first visit to Cornwall must be pronounced a success for, in spite of all setbacks, he was able to write (1777, Sept. 20) just before his return to Birmingham: "The voice of the country seems to be at present in our favour and I hope will be much more so when the engine (i.e. Chacewater) getts on its whole load which will be by Tuesday next. So soon as that is done, I shall set out for home."

On arrival, he found a large amount of work awaiting him and this occupied him till the end of the year. Next year, 1778, was one of unexampled physical and mental exertion, as evidenced by his remark to Boulton on February 2nd: "I fancy I must be cut in pieces and a portion sent to every tribe in Israel." In May, accompanied by his wife, he again set out for Cornwall. They rented a house at Plain-an-guary on the outskirts of Redruth in a convenient position for most of the mines (see Pl. IX (a)).

Watt was at once immersed in work, for it has to be remembered that, besides going out on the mines to superintend the erection of the engines, he was busy drawing arrangements and details for engines in hand at Birmingham. Nothing went right; there were mistakes in castings, defects in materials, defects in workmanship and delays in delivery. "I am very sensible that in ye mult[itude] of things I have to think off & the vexations I sometimes meet with, I am not so accurate in giving out orders as I might be." Yet what could be expected with such a new enterprise? The truth is Watt was over-burdened with detail. Besides this, there were certain constructional details that were far from satisfactory, and one of these was piston packing. Watt was still experimenting with it, as shown in a letter of Aug. 3rd, 1778.

The packing consisted merely of short wedge-shaped segments of anti-friction metal with overlapping joints placed next the cylinder wall; elasticity was supplied by oakum behind them. This is the earliest instance known to the author of the application of metal to piston packing, although not of course as we know it to-day.

Other letters follow, full of engine and pump troubles. We select as typical that of Aug. 13th, 1778:

"Chacewater engine wore a very gloomy aspect" and "went sluggishly". Watt made a thorough inspection of it, thinking "perhaps a board, a bunch of oakam, somebody's hat or coat, had been left in ye cylinder & had come into ye nozzle (i.e. the steam valve)"; instead he found that part of the latter had come unsoldered and when it was repaired they "went to work immediately". Another trying job was to make tight joints, a perennial job with the engineer. Watt at this date used straw board smeared with thin putty.

Boulton gave his views on this and the subject of

piston packing, to which Watt's rejoinder (1778, Sept. 12) was:

I had discovered all you mention in relation to piston leads before I got yours; but imprimis a mixture of Lead & tin is much worse than lead for its melting point being abt 300ᵈ it grows so soft as to tear itself in pieces by yᵉ heat which lead does not do so badly—2dly the too great inclination of yᵉ cones is true but its effect may be lessened by the slope of yᵉ upper side no pasteboard

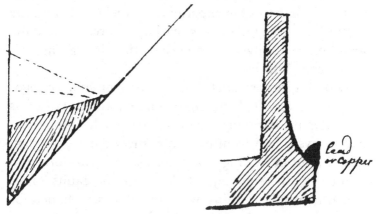

should be put under yᵉ leads but laid upon the bare cone—They may be cast in long slips of copper with 1 oz tin to pound and afterwards bent to lay in place—The new piston I propose is of this fashion for here the roping will come always into a narrower & narrower place and by virtue of yᵉ semi elliptical figure will end at right angles to cylinder and have no inclination to sink in ye oackum though no lead were used but I think it will be best to have a small slip about $\frac{1}{2}$ inch, which will [be] pressed by the oackum in proportion to its upper slope. The oackum & roping should be quite white i.e. free from any tarr & then very little Grease is necessary, for yᵉ article cannot be afforded particularly here where yᵉ Engine men eat it Adieu—Upper leads of piston should be short and deep.

It is characteristic of Watt that if anyone suggested any improvement to him, he reckoned he had already thought of it; that sort of thing is exasperating but in his case there is not much doubt that it was true. The touch about the enginemen eating the grease reminds one of similar trouble experienced in Russia before the Great War owing to moujiks eating the grease out of the axle boxes of railway wagons and thereby causing the bearings to run hot.

Looming larger than troubles with engines were negotiations in respect of new engines now wanted. Watt was no better at a bargain than he had been in his Monkland canal days. The adventurers were shrewd men, rough in speech, but of sterling character as Watt was forced to admit (1778, Aug. 29): "With all the faults of ye Cornish people, I think we have a better chance for tolerable honesty here than elsewhere, as their meetings being public they will not chuse to expose themselves any farther than strict dealings may justify and besides there are generally too many to cabal." He longed for Boulton, a tower of strength in such matters, to come into Cornwall, for as Watt says later in the letter above quoted: "As Whⁱ Union will probably be going next Monday or tuesday, I think you should be turning your thoughts to a journey here and the final settleing of all the agreements in this county—Chacewater engine is our capital card for should it succeed in forking the mine, all doubts will be removed."

Boulton contrived at the end of October to pay the visit to Cornwall for which Watt pleaded. He went round the mines and succeeded in arranging definite terms, not only for the use of the engines already erected such as those of Chacewater and Ting-tang, but also for such as were in prospect.

To Watt, whose horizon did not extend beyond Cornwall and the competence that could be drawn from that limited field, it seemed that the harvest of profits had begun. How different it appeared to Boulton! Owing to his commitments in his businesses— and he had many apart from the steam-engine—he found himself greatly in need of ready money. It is difficult for us nowadays to realise that at the end of the eighteenth century there was no way of raising capital for an enterprise except by obtaining a charter or by taking in a partner. Even if a desirable one could be found, the liability involved was unlimited and was one not readily faced. It was possible to raise a short-period loan by way of bills or mortgages and lastly recourse could be had to bankers, but this only afforded temporary relief. Nowadays Boulton could have floated a company, and with his reputation we can be sure that any capital he wanted would have been over-subscribed many times.

The favourable reception of the engines in Cornwall was a godsend to Boulton as it enabled him to offer to the bankers, as security, the profits of the engines already erected. To this course Watt's consent was necessary and was most reluctantly, we might almost say despairingly, given. Boulton succeeded in getting a sum of £2000 from Elliot & Praed, bankers of Truro, and this was not all, for he obtained a further large sum from Lowe, Vere & Co., bankers of London. These debts were a nightmare to Watt, who never ceased groaning under the burden, anticipating ruin for himself and destitution for his family. The language he used was pessimistic in the extreme, in fact it bordered on the ludicrous and we spare the reader by not quoting it.

Boulton's difficulties were serious, it may be admitted, but only temporary, for Soho Manufactory was full of orders and it was now clear that the engine business was going to be remunerative. It is a tribute to Boulton's fine character that he never lost patience with Watt; however, we have gone ahead too fast with our story.

Watt remained behind in Cornwall and did not return to Birmingham with Mrs Watt till early in the New Year, only to find as before an accumulation of work awaiting him. Fortunately for him, relief was at hand, at any rate for the outside work, in the person of William Murdock, a stalwart who was destined to become the right-hand man of the firm in Cornwall and later a tower of strength at Soho. The story of Murdock's interview with and engagement by Boulton in 1777 is told by Smiles. The young Scot, bashful and nervous in the presence of the great man, kept twirling his hat, thereby drawing Boulton's attention to it. Thinking it looked peculiar, he asked Murdock what it was made of.

PLATE X. WILLIAM MURDOCK
From the oil painting by Graham Gilbert. Published by permission of the
Museum and Art Gallery Committee of the Corporation of Birmingham

"Timmer, sir! I turned it mysel on a bit lathey of my own making." The only lathe that would do such a job was an oval lathe, the construction of which was evidence of mechanical knowledge and of skill of a high order. That was enough for Boulton; within a fortnight Murdock was engaged, and in 1779 he was sent to Cornwall as one of the engine erectors. There he came to the front quickly, for he was intelligent, resourceful and temperate in the matter of strong drink; the latter characteristic was not too frequently found in peer or peasant in those days! Boulton described him as "the most active man and best engine erector I ever saw".

Murdock was of striking appearance as our illustration (Pl. X) shows; he was an inventor of no mean capacity as is revealed in succeeding pages, and the best proof of his value is that the firm kept him in their employ until the day of his death.

Watt's repeated absences from Birmingham on business, and the necessity to keep track of his correspondence, entailed making copies of letters—drudgery that he found most irksome. Stimulated by the need for some mechanical aid to relieve this drudgery, he was led to think of offsetting the writing by pressure against an unsized damped sheet of paper, when the writing could be read by looking at it from the *reverse* side. Success depends entirely upon the ink used, which has to be mixed with mucilage to make it offset clearly. Although not the first to observe that the addition of sugar or gum to tannin ink will enable it to offset, yet Watt must be credited with being the first to make a useful application

8-2

of the property. He experimented assiduously during May and June to perfect the process and made the first announcement of the invention to his partner in the following words (1779, June 28):

I send you enclosed some of Mr Nobody's draughts with authentic copies of them; You will observe that that which appears full black will not preserve even the colour it has but will grow browner by keeping. The paler kind are nearly if not quite fully as white as the paper was before the operation and will stay so.

I will be much obliged to you to procure me a quire or two of the most evenly & whitest *unsized* cambric paper, and also specimens of the cheaper kinds with their wholesale prices. If you apply at Curtis's [you] may perhaps be better served as we have dealt with him before. It is absolutely necessary that there be no size in the paper which [you] may know by touching it with a wet finger.

The copies will continue to grow blacker than they were before copying and as far as I can judge not in the least defaced.

Watt completed the invention by designing a press to take off the copies. Since at this time he used only single sheets of paper, he utilised the roller press although he envisaged also the screw-down press which was found preferable when much later the press-copying book came into use. Both presses are described in the specification of the patent (1780, Feb. 14, No. 1244) that he took out for this invention. On March 20th, a new firm —James Watt & Co.—the partners in which were Watt, Boulton and James Keir, was formed to exploit the invention. Boulton with his usual enthusiasm took a press to London, showed it to members of both Houses of Parliament, to bankers, and to the frequenters of coffee-houses, to such good effect that by the end of the first year 150 presses were sold. What these were like can be judged from our illustration (Pl. VII (*b*)), which is taken from a press in the Science Museum. The invention was received at first with alarm from the supposed danger of forgery, but its inherent merit gradually made it almost universal as a means of copying correspondence till within recent times when the typewriter carbon copy has superseded it.

From 1780 onwards, press-copying was used exclusively by the group of firms at Soho, not only for correspondence but also for working drawings; since the latter had to be on thick and therefore opaque paper, they had to be read on the front. Naturally the drawing was reversed so that the figured dimensions and the colouring had to be added after copying. Many thousands of these drawings marked "Reverse" are to be found in the Boulton and Watt Collection at Birmingham. Two of our illustrations (pp. 154 and 180) incidentally exemplify this practice.

The ink, as we have said, was all important, and to ensure success it was made and supplied by the firm in powder form. How important may be judged by a recipe for it in a letter still in existence (1782, Jan. 15) from Watt to his partner Keir, a letter remarkable alike for the intimate knowledge it displays of the properties of the ingredients as for the details of the manufacture.

Before leaving the subject of press-copying, it may be

mentioned that in 1794, when the patent was about to run out, James Watt junior had just begun his connection with Soho and it was obvious that the business of making presses would very quickly be taken out of their hands unless the firm made some vigorous move. He, therefore, designed a neat portable machine on the lines of a folding writing desk. He printed "Directions" for using it in English, French and German and appointed agents in several places. Such desks are still to be found; one such is in the Boulton and Watt Collection, Birmingham.

Watt used the roller press to within a few days of his death and his letter books in which he pasted the flimsies are still preserved at Doldowlod. Incidentally a specimen of his press-copying is reproduced on p. 195. We like to reflect occasionally that Watt's press was the forerunner of those adjuncts to commercial life with which to-day every office is replete.

Watt was all for shortening labour and he found the arithmetical calculations he had to make just as irksome as he had found letter-copying, so that he now invoked mechanical aids to arithmetic; these he found in the slide rule. In a letter (1779, Oct. 30) to Boulton who was then in town, Watt wrote: "If you have time, I wish you would call at Nairne's and gett one of Mountain's sliding rules. I believe they are about 2 feet long, the longer the better, but if any of (them) have the double radius or line of squares next one side of the slide, they will answer our operations best; but bring one at any rate. . . ." Watt did not, as some have thought, design a rule for engineers' use; he merely chose the most suitable pattern to be had on the market; but we can say that he was the first to employ the slide rule in engineering calculations. A slide rule, believed to be one of Nairne's make, is preserved in the Watt Workshop at South Kensington.

It might be supposed from what has been said of Watt's intense mental application and close attention to business that he had no recreations beyond the one we have mentioned, viz. reading, yet such was not the case. It is true that physical exercise at the time of which we speak was largely unnecessary, because most persons got sufficient of it in their daily work. But there was a social side to Watt's life at Birmingham and this may best be exemplified by his membership of the Lunar Society. In those days there was no gadding up to London to attend meetings, and in consequence in many of our large towns such as Manchester and Birmingham, literary and scientific men came together in groups for intercourse. The Lunar Society was one of the most famous of these bodies and was so called because its meetings were held at the time of the full moon in order that Members could get home afterwards without difficulty. Watt had been introduced to the Society on one of his early visits to Birmingham; he now had the pleasure of becoming a permanent member. What the meetings were like can best be visualised from Dr Darwin's note of apology to Boulton on one occasion: "Lord! what inventions, what wit, what rhetoric metaphysical, mechanical and pyrotechnical will be on the wing, bandied like a shuttle cock from one to another of your troop of philosophers." At one time or another the Society, or "Lunatics" as they were irreverently called, had many members besides those just mentioned: Dr Small, James Keir, Thomas Day, Richard Lovell Edgeworth, Samuel Galton, Dr Withering, and Josiah Wedgwood. Then again, since each member could bring a guest, many a distinguished visitor found his way to the assemblies. We may mention John Smeaton, Sir Joseph Banks, Sir William Herschel, Dr J. A. Solander and J. A. de Luc.

Such were the recreations of our philosophers and such the happy *milieu* in which Watt expanded socially and intellectually.

By the end of 1780 Watt had been relieved from pecuniary worry, although not Boulton, for the firm had at last begun to show a profit. Up to December 31st, the partners had received from Cornwall nearly £2600 by way of premiums. This state of affairs is alluded to by Watt in a jocular letter to Gilbert Hamilton (1781, April 9) thus: "Our general expenses have hitherto been very great so that the business never paid its way before 1780 when I guess it got about £2000 which was all swallowed up by original sin as more must be."

The very success of the firm, however, excited envy, particularly among the miners in Cornwall. The younger generation did not realise how much the mining interest owed to Watt's engine, for there was scarcely a common engine left with which to compare it. All that these miners knew was that Boulton and Watt were holding them to ransom, as they considered, and extracting money from them that they thought should be in their own pockets. The first rumblings of the storm that was brewing reached the partners towards the end of 1780. This occasioned a magnificent outburst from Watt, quite in the vein of his covenanting forefathers, and we cannot refrain from quoting from it (1780, Oct. 31):

They charge us with establishing a monopoly, but if a monopoly, it is one by means of which their mines are made more productive than ever they were before. Have we not given over to them two-thirds of the advantages derivable from its use in the saving of fuel, and reserved only one-third to ourselves, though even that has been further reduced to meet the pressure of the times? They say it is inconvenient for the mining interest to be burdened with the payment of engine dues; just as it is inconvenient for the person who wishes to get at my purse that I should keep my breeches pocket buttoned. It is doubtless also very inconvenient for the man who wishes to get a slice of the squire's land that there should be a law tying it up by an entail. Yet the squire's land has not been so much of his own making as the condensing engine has been of mine. He has only passively inherited his property, while the invention has been the product of my own active labour and of God knows how much anguish of mind and body.... Why don't they petition Parliament to take Sir Francis Bassett's mines from him? He acknowledges that he has derived great profits from using our engines, which is more than we can say of our invention; for it appears by our books that Cornwall has hitherto eaten up all the profits we have drawn from it, as well as all that we have got from other places, and a good sum of our own money into the bargain. We have no power to compel anybody to erect our engines. What then will Parliament say to any man who comes there to complain of a grievance he can avoid?

This was the first threat to their patent monopoly; we shall have to explain later how this opposition developed and in what litigation the firm was eventually involved. For the time being, the opposition seemed to die down and this, coupled with his satisfactory financial position, engendered, we feel sure, the state of mind that favoured the next outburst of Watt's powers—that of the perfecting of the rotative engine with the inventions depending thereon.

Watt's home, Regent's Place, Harper's Hill, Birmingham, 1777
From Smiles's *Boulton and Watt*, 1865

CHAPTER VI

PARTNERSHIP WITH MATTHEW BOULTON. THE ROTATIVE ENGINE, 1781–1790

Demands of industry for mill engines. The crank and its substitutes. Iron cement. Expansive working. Hornblower's infringement. Parallel motion. Testing materials. Horse-power. Albion Mill. Bleaching by chlorine. The Governor. Boilers. Smoke consuming. Decimal system.

IN his forty-fifth year and in the hey-dey of his intellectual powers, Watt now entered upon the most brilliantly inventive period of his career, that of the development of the rotative engine capable of unlimited application in industry. Up till this time the steam-engine had been purely and simply reciprocating, its only application being pumping water. Now it was about to fill a new rôle—that of turning all kinds of millwork in any locality and to any desired extent. No one knew better than Boulton the pressing need of industry for more power and no one had deeper insight into what future requirements would be. He had talked rotative engines for two years at least and, in June 1781, he felt he must force the hand of Watt who was dreaming only of pumping engines and of another Cornwall. Mr and Mrs Watt had at this juncture arrived in Cornwall and were entering into residence of Cosgarne House. When it became pretty clear that one or other of the partners would have to spend some time each year in

Cornwall, Boulton came to the conclusion that it would be desirable to maintain an establishment there where either partner might reside as occasion required. They found what they wanted at Cosgarne in the Gwennap Valley, which although little more than a mile from the United Mines, is yet secluded from the mining area. It was and remains to-day as described by Watt "a most delightful place, a neat roomy house with sash windows double breadth the front to the south". Our illustration (Pl. IX (b), p. 109) bears out these encomiums. But to resume, Boulton's attitude is clearly shown in the following extracts from a letter of his (1781, June 21):

"The people in London, Manchester and Birmingham are *steam mill mad*. I don't mean to hurry you but I think in the course of a month or two, we should determine to take out a patent for certain methods of producing rotative motion from...the fire engine" and again "There is no other Cornwall to be found, and the most likely line for the consumption of our engines is the application of them to mills which is certainly an extensive field."

It is a common misapprehension to suppose that the steam-engine created the new industrial order that was then arising; it did give industry an enormous impetus and was the most important single factor in helping it on but industry was already in being and would have flourished if the steam-engine had never arrived.

To make the engine turn millwork it was only necessary to fix to the shaft a crank and a flywheel to carry the crank over the dead centre and to apply a connecting

rod from the beam to the crank. To us to-day nothing
seems more obvious and the application demands no
more than a moment's consideration. It was far otherwise then. In the first place the stroke of the engine was
neither uniform, nor invariable in length. With a crank
and a short stroke, it was imagined that the engine
would stop or turn back, and with a long stroke that
something would have to give way. It had not dawned
upon anyone, even Watt, that a flywheel and a crank
could successfully *control* the movements of the reciprocating parts. In the second place, the engine was only
single-acting and to get it to make the return stroke, a
heavy weight sufficient to balance half the effort of the
piston would have to be placed somewhere, either on
the end of the beam or on the connecting rod, unless
entirely different means of equalising the effort of the
piston could be found. The problem is one that was
present in the minds of a great number of engineers at
this time.

The story that has grown up about Watt's connection
with the crank is that he succeeded in applying it to the
steam-engine but that his ideas were communicated, by
the treachery of one of the Soho workmen, to a rival
engineer who took out a patent for the crank and successfully deprived Boulton and Watt of the use of it till the
patent expired. The facts are more prosaic. Watt never
claimed the application of the crank to the steam-engine,
indeed he did not consider it a patentable one, for as he
said "applying it to the engine was like taking a knife to
cut cheese which had been made to cut bread". What he

invented, and what was stolen from him, was the combination of a crank with revolving counter-weights, an arrangement which if it ever materialised was soon abandoned.

The story of what happened is somewhat involved, but we shall tell it as shortly as possible. James Pickard of Snow Hill, Birmingham, in 1779 wanted an engine to drive his flour mill there and he employed Matthew Wasborough, an engineer of Bristol, to make it. To convert its reciprocating into rotative motion, Wasborough fitted the engine with a pawl and ratchet arrangement, which he had patented that same year. This mechanism was an old one really, often tried but always discarded because of its clumsiness and liability to break down. In the case of Wasborough's engine, the result was no different to previous attempts and at the end of a year the ratchet and pawl was replaced by a crank and connecting rod. Seeing that the engine already had a flywheel, the problem was solved. Pickard and Wasborough deserve credit in that they had the courage to apply the crank and take the risk of a breakdown which was what everyone expected to happen. It must not be thought that Wasborough realised at first that it was the flywheel that had saved the situation. By this time Pickard had obtained a patent [1780, Aug. 23], the specification of which describes the combination of a crank with a wheel having on the rim weights revolving at twice the speed of the crank shaft, but it does not mention a flywheel. It is this mechanism that there is evidence had been divulged by the tittle-tattle of a Soho workman; it is known that Watt had thought about this mechanism and had even made a model of it a year before Pickard took out his patent. Yet why did Watt not patent it himself? The excuse offered is that the partners were so much employed in Cornwall, but it is more likely that Watt felt it was an unsatisfactory device and therefore best left alone. Nevertheless, Watt was very angry with Pickard and talked of contesting his patent but evidently thought better of it, reflecting probably that it might only encourage reprisals against his own patents. Watt took no action and as it was generally assumed, wrongly as we believe, that Pickard's patent covered every application of the crank to the engine, Boulton and Watt avoided the use of the crank till the patent expired.

Wasborough and his associates entered into negotiations with the firm to obtain a licence to use the separate condenser in return for a licence to use the crank. To this proposal Watt was inflexibly opposed so that nothing came of the negotiations. Instead Watt set to work to devise substitutes for the crank. He had already schemed the one that Pickard patented and others were: a two-cylinder engine with cranks at an angle and counterweights on the shaft; the ladder connecting rod; the internally-geared connecting rod; the swashplate or crown cam motion and the sun-and-planet gear. These and other mechanisms he included in the patent that he took out on October 25th, 1781, for "Certain new methods of applying the vibrating or reciprocating motion of steam or fire engines to produce a continued rotative or circular motion round an axis or centre and thereby to give motion to the wheels of mills or other machines".

Of these the only one that came into use was the sun-and-planet or epicyclic gear, so that we shall not weary the reader by describing the rest but we may say that models of them are to be seen in the Watt Collection at the Science Museum. The first mention of the sun-and-planet gear is in a letter to Boulton from Watt in Cornwall, dated January 3rd, 1782, who describes it as follows:

I wrote you on the 31st since which I have tried a model of one of my old plans of rotative engines revived and executed by W. M(urdock) and which merits being included in the specification as a fifth method, for which purpose I shall send a drawing and description next post. It has the singular property of going twice round for each stroke of the engine and may be made to go oftener round if required without additional machinery. The wheel A is fixed on the end of an axis which carries a fly. The wheel B is fixed fast to the connecting rod from the working beam and cannot turn on its axis, and is confined by some means so as always to keep in contact with the wheel A. Consequently by the action of the engine it goes round it and causes it to revolve on its axis and if the wheels are equal in the number of their teeth, A will make two revolutions while B goes once round it.

The description will be still clearer if reference is made to Pl. XI (*a*).

In fulfilment of his promise, Watt sent the drawing

PLATE XI (*a*). MODEL OF SUN AND PLANET GEAR, 1781
Courtesy of the Science Museum

PLATE XI (*b*). EXPERIMENTAL MODEL OF
ENGINE BEAM, *c.* 1786
Courtesy of the Science Museum

with this note (1782, Jan. 7): "I wrote you on Saturday with drawings of the 5th method of rotatives; and inclosed I send the complete specification of that method."

It will be observed that Watt says that the sun-and-planet gear was "revived" by Murdock; there is evidence that it was invented by him independently. However that may be it was he who suggested it to Watt, with the result that the latter inserted it into the specification at the last moment.

The firm used the sun-and-planet gear on their engines during the whole period of the patent and even subsequently to 1794 when they were free to use the crank; in fact the gear was supplied as late as 1802, so hard is it for convention to say goodbye to the obsolete. A representation of the state of development of the rotative engine when the crank came into common use is shown on a subsequent page (see Fig. 11, p. 172).

Watt had trouble from the beginning in making steam-tight joints, and he frequently turned over the matter in his mind. Hence the inception of another valuable invention, that of the well-known iron cement, in which again there is evidence that Murdock had some share. This cement, so much used by engineers at one time, is composed of a mixture of iron borings, sal-ammoniac and sulphur. The date of Watt's first experiments in this direction is fixed by a letter to Boulton from Cornwall (1782, Apr. 10) in these words:

—am trying some experiments on a new cement for Joints, as I have lost faith in putty which always fails in the long run in all the vacuum Joints owing to the repeated exhaustions and repletions, I have several sorts in view one, a fine powder of Iron in a metallic state mixed with substances which may dispose it to rust and some mucilaginous matter which may give it elasticity and keep it in place till it rusts, and a new substance which is neither soluble in ∇ [i.e. water] $\sqrt{}$ [i.e. acids] nor Oils but can be softened by water to the consistence of Caoutchou & which when mixed with earthy substances becomes as hard as stone, and can be had cheap, being the product of an English vegetable. What continued heat may do to it experience must determine. In other respects it would make a hole of an inch wide quite tight in one minute after applied and adheres to any dry substance most viciously. Caustic alcalies or acids destroy it or weaken it.

What "the product of an English vegetable" was we do not know but it could not be sal-ammoniac; nor is there a specific mention of sulphur. Watt in an account written in 1814 for Brewster says that Murdock independently "made a cement of iron borings and sal-ammoniac without the sulphur. But the latter gives the valuable property of making the cement set immediately". The impression one gains from the facts adduced is that the two ingenious men were experimenting simultaneously and the cement that we know was the outcome of their having put their heads together, a joint invention, if we may be excused the pun. From 1784 onwards, the cement was made at Soho in quantities and was sent out to every job; naturally it became known outside Soho and was soon in general use by engineers.

We envisage Watt by this time in comfortable circum-

stances, his health improved and his financial worries
abated, but he was still tied down to the desk and drawing board. Boulton had urged him continually to get
help, but Watt was one of those rather trying persons
who want to do everything themselves and are not
content with the efforts of others. He had had two
assistants, but he did not approve of either.

At length in 1782, John Southern, a son of Thomas
Southern of Wensley near Wirksworth, Derbyshire,
then twenty-four years of age, was introduced by
Boulton. A good deal of the rigid Presbyterian in Watt
is revealed in his response to Boulton's good offices: "If
you have a notion that young Southern would be
sufficiently sedate, would come to us for a reasonable
sum annually, and would engage for a sufficient time, I
should be very glad to engage him for a drawer, provided he gives bond to give up music, otherwise I am
sure he will do no good, it being the source of idleness."

Southern was engaged and, in spite of his musical
proclivities which we do not believe he sacrificed, became Watt's right-hand man. Southern spent the rest
of his life in the service of the firm, having been admitted
a partner in 1810, and died in 1815.

Now that Watt had reliable assistance on the outside
erection work and in the drawing office, and had disposed of the equivalents of the crank, he was free to turn
his attention to further conquests. A patent quickly
followed (1782, March 12, No. 1321) for "certain new
improvements upon steam or fire engines for raising
water, and other mechanical purposes, and certain new

pieces of mechanism applicable to the same". In the specification, in the first place, he described expansive working of steam. He had had this idea in his mind since 1769.

The advantage of working steam expansively is that the economy is effected by filling the cylinder only partly with steam, then cutting it off and allowing the remainder of the stroke to be completed by the gradually diminishing pressure of the steam as it expands. To carry the idea into practice, required alterations in the steam and exhaust valves. In his specification of 1782 he gives a clear description of the advantages of expansive working and, further, supplies a remarkable diagram of what goes on in the cylinder when the steam is cut off at one quarter of the stroke. The diagram is not based on actual observation but upon Boyle's law of the expansion of gases. Watt had tried expansive working in 1777 with the engine at Soho known as "Beelzebub", afterwards rebuilt and known as "Old Bess", now one of the treasured engines at the Science Museum.

Watt lost interest in the idea because he could see that there was very little saving to be made with steam such as he employed that was initially only of atmospheric pressure. After Watt's time, when steam pressures had been increased materially, expansive working had important influence on steam-engine economy. As it was he patented quite a number of ways in which the varying power due to expansion could be equalised.

Watt was not the only one to conceive the idea of using steam expansively. Jonathan Hornblower the younger also had the idea, but his plan was to use steam successively in two cylinders—nothing more or less than the compound engine principle. Hornblower is said to have made a model of his engine in 1776, but it was not revived till 1781 when he took out a patent for it. His first engine was built for Radstock Colliery near Bath; one cylinder was 19 in. diam. by 6 ft. stroke and the other 24 in. diam. by 8 ft. stroke. Rumours of the new engine quickly reached Watt's ears, and when he learnt its true character he became thoroughly alarmed. The Radstock engine, after the usual troubles, was got to work in the autumn of 1782, and performed satisfactorily. Watt's state of mind is revealed in the following letter (1782, Sept. 28):

I received your's of the 23^d which though no more than I had reason to expect gives pain—You ask my advice—In the first place I think it too late now to apply to the owners of Radstoke except by an Attorney. It might suffer a bad construction, besides we should not warn a man that we mean to break his head, lest he put on a helmet....

Secondly at present I think there is no alternative but as soon as prudent to commence a suit against all the *Horn*blowers and their Radstoke friends....

However much I am vexed I am rouzed and shall prepare myself to meet the worst and not lie down to have my throat cut—I beg you would summon up your resolution & not lose the battle before you fight it.

Watt's decision to die in the ditch, so to speak, was really ludicrous and even more so was his advice to Boulton to "summon up your resolution". It was rather a psychological effort to summon up his own courage. The partners informed the proprietors of Radstock that Hornblower's engine was an infringement of theirs, although Hornblower argued that he was condensing the steam in his second cylinder and was not using the separate condenser.

However, Hornblower's engine made little progress, for with the pressures then used it was no more economical, and it was more costly to build than Watt's engine. One was put up at Penryn in 1784, another at Tincroft in 1791, and in all about nine or ten. Boulton and Watt demanded premiums on these engines and the owners paid up eventually, rather than face litigation. The position was that Hornblower had a patent which he could not work so long as Watt's was in force and that the partners would not deviate from their policy never to grant licences to other firms to use the separate condenser. Hornblower, seeing that his patent would expire in 1795, applied in 1792 for an extension of his patent, but Boulton and Watt opposed him so successfully that his Bill did not pass into law.

But we have strayed from our text, which was the specification of the patent of 1782. Watt included besides expansive working, the double-acting engine, the advantage of which is that from the same sized cylinder double the power can be got. This idea had been in his mind since 1775 and he had in fact submitted a drawing of this engine to Parliament; our illustration (Fig. 7) shows his ideas at that date. Since in a double-acting

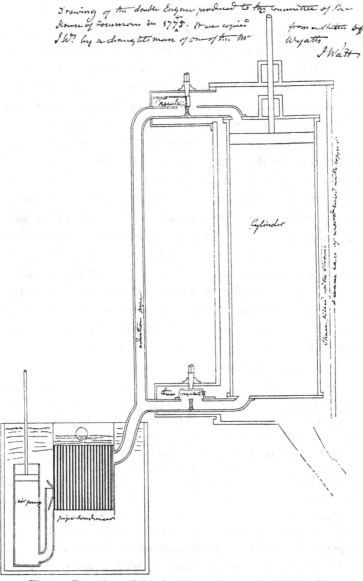

Fig. 7. Drawing of the double-acting engine laid before Parliament, 1775

Doldowlod Papers

engine the piston pushes as well as pulls, the flexible connection of the piston to the beam was no longer admissible. Hence Watt had to scheme some other means of connection. One of his mechanisms, included in the specification, was a rack piston rod gearing with a sector on the beam. This was actually applied to an engine but it was clumsy and some other mechanism was called for. To this task Watt now applied his mind. It may reasonably be asked why he did not adopt what seems to us to-day so simple, viz. a crosshead and guides for the piston rod. As a matter of fact he did include guides in the patent of 1784, but we have to remember those were the days before planing machines and to produce plane surfaces by chipping and filing by hand was far too costly for an engine of any size. As plane surfaces were out of the question, Watt turned to systems of linkages and hit upon what we know as the parallel motion. It is best described in Watt's own words (1784, June 30):

I have started a new hare! I have got a glimpse of a method of causing a piston rod to move up & down perpendicularly by only fixing to it a piece of iron upon the beam, without chains or perpendicular guides or untowardly frictions, arch heads or other pieces of clumsiness, by which contrivance it answers fully to expectation...about five feet in the height of the house may be saved in 8 feet strokes....I have only tried it in a slight model yet, so cannot build upon it, though I think it a very probable thing to succeed, and one of the most ingenious, simple pieces of mechanism I have contrived.

In little over a week (July 11th) he had tried it on a larger scale.

I have made a very large model of the new substitute for racks & sectors which seems to bid fair to answer. The rod goes up & down in a perp^{lr} line without racks, chains or guides. It is a perpend^{lr} motion derived from a combination of motions about centres, very simple, has very little friction, has nothing standing higher than the back of the beam and requires the centre of the beam to be only half the stroke of the engine higher than the top of the piston rod when at lowest & has no inclination to pull the piston rod either one way or another except straight up & down...however, don't pride yourself on it, it is not fairly tried yet & may have unknown faults.

This is the three-bar motion. The "motions about centres" are those of the engine beam and the radius rod. These two are joined by a link and there is a certain point in the link which has practically no sensible variation from a straight line when the linkage is moved. Watt investigated the properties of the linkage very carefully, by actual models be it noted, and found it fulfilled his hopes. This, with two other methods of guiding the piston rod, were protected by the patent of 1784 (Aug. 24th). The three-bar motion was actually applied to two engines, but it had a drawback from the practical point of view, i.e. it required the radius rod to stand out beyond the beam, and this meant that the engine house must be enlarged to that extent to take it.

Within a few months, Watt had added a pantograph extension to the three-bar motion, thus arriving at the elegant parallel motion. Our illustration of this motion (Fig. 8) taken from one of the firm's drawings after the construction had been standardised, shows what it looks like applied to the engine, and this should be studied with the diagram (Fig. 9) taken from the drawing and to the same scale. The three-bar motion is picked out in full lines and the pantograph extension in dotted lines.

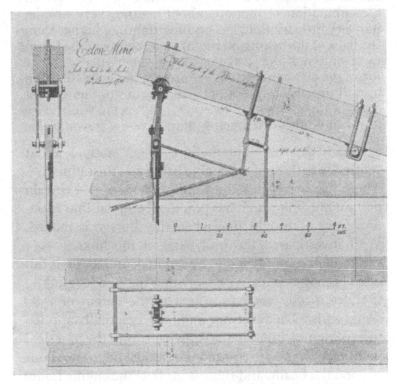

Fig. 8. Parallel motion of Ecton Mine engine, 1788
Boulton and Watt Collection. Courtesy of Birmingham
Reference Library

The improvement, it will be observed, consists in arranging the lay-out so that the piston rod instead of being attached to the straight line point of the three-bar motion is transferred to the corresponding point in the pantograph extension; thus the mechanism is accommodated snugly within the limits of the ordinary engine house. The diagram shows also in full lines the complete closed curve that would be described if the movement of the linkage was unconstrained; only that portion *AB* of the curve which does not deviate sensibly from a straight line is made use of in the engine.

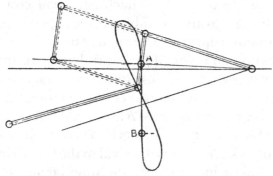

Fig. 9. Parallel motion, based on preceding illustration

No wonder Watt was proud of the invention and could write in later life (1808) to his son: "Though I am not over anxious after fame, yet I am more proud of the parallel motion than of any other mechanical invention I have ever made." There speaks the true craftsman!

We have not yet done with this omnibus patent of 1782, for in it were included various applications of the double-acting engines, e.g. to wheeled carriages and to tilt hammers with or without the condenser. There does not seem to be any doubt that the claim with regard to steam carriages was put in because Murdock had taken up the subject that summer in his spare time and was engrossed in it. Watt and Boulton not only wanted to keep Murdock quiet but wished to exclude others, for numbers of persons were experimenting with steam carriages. It is abundantly clear that the partners had no intention of going into the matter; this is proved by a letter of quite extraordinary length from Watt to Boulton (1784, Aug. 31) winding up with this dictum: "You will see that...the machine...will cost much time to bring it to any tolerable degree of perfection and for me to interrupt the career of our business to bestow my attention on it would be imprudent." In the light of the experience of subsequent inventors of steam carriages, the conclusion was eminently sound, but one does not see any reason why Watt should have been such a dog in the manger to Murdock. He was valuable and the partners knew he was too loyal to the firm to leave it, so they talked him over for the time being. He kept hankering after steam carriages, however, as is shown by a letter of two years later date from Watt to Boulton (1786, Sept. 12). As far as we can find out, the alliance of Watt with the Deity adumbrated in this letter did not "work a miracle in favour of these carriages" and their appearance on the roads was delayed for nearly half a century. The letter is as follows:

I am extreamly sorry that W. M[urdock] still busys himself with the St^m Carriage....I have still the same opinions concerning it that I had, but to prevent as much as possible more fruitless argument about it, I have one of some size under hand & am resolved to try if God will work a miracle in favour of these carriages.

The first rotative engine to be sent out was to the order of Wilkinson, who was always first on the list for anything new, and was supplied at the end of 1783 to Bradley Ironworks, where it worked a tilt hammer. Watt was, however, too deeply engrossed in the double-acting engine mentioned above to be able to give the attention he would have wished to the rotative engine, so that the next one was not finished till 1784. A few years later it had become standardised and the double-acting engine with parallel motion, sun-and-planet gear, valves worked from the beam, timber framing beam and connecting rod, was being made for all and sundry. We shall have occasion to revert to this engine after we have touched upon other activities of the many-sided Watt, like wefts through the warp of his life.

Watt's mind was essentially deductive; it was instinctive with him to "try all things". We cannot resist instances concerned with testing materials. The source of our information is a "Blotting and Calculation Book 1782 to 1783" preserved in the Boulton and Watt Collection. This book is evidently a record of data that would be useful in designing engines and machinery. One of the items is an "Experiment on the stiffness of iron Augt 1783", the purpose of which appears to have been to simulate a piece of a machine exposed to repeated shocks. Other experiments recorded are bending and breaking tests of bars of cast iron, wrought iron, oak and deal.

Another instance of Watt's resort to experiment and his deductions therefrom is afforded by an accident that happened in 1786 to the gudgeon or crank shaft bearing of one of their engines. The proprietors condemned the iron of which it was made as bad and wanted Boulton and Watt to pay for its replacement; to judge by what is stated to have happened to the engine, no material on earth would have stood the shock to which the gudgeon had been subjected. This was Watt's opinion and his report to Boulton of the result of his experiment in a long letter (1786, March 22) is so informative that we give it in full:

Yesterday I crept out to Soho & tried the Iron of F. Scott's Gudgeon, when laid with the angular side downwards between two sup- ports about 5 inches asunder. We could not break it until it was notched cold on the ridge pretty deep & after notching it stood many severe blows from one of our strongest men with a hammer of near 40^{lb} weight before it broke but when it broke it was all at once & shewed the same chrystallized grain as in y^e org^l [i.e. original] fracture.

I then splitt a piece of the Edge of it & caused draw it out into a rod of $\frac{1}{2}$ inch square and abt 12 inch long this rod bore being quite bent double in the middle, but broke in attempting to straighten it again. I then gave one of the pieces a slight notch & put it in the vice, it broke quite short without bending, & showed a cold short grain. I served the other end the same way & it bent somewhat before it broke & shewed some tough & a smaller grain.

In the opinion of our Smiths a barr of good comm^n Swedish w^d have broke as easily (though of same size) as this did in the mass, though it w^d probably have been tougher in the small piece. On the whole it does not appear that the Iron was remarkably bad, but such accidents as broke the fly & tore up the plummer blocks w^d demolish almost any hard Iron & soft Iron is not fitt for Gudgeons. I know not what country Iron it is but W^n [i.e. Wilkinson] used to get his Iron for Bersham from Cumberland Charcoal Iron throughout. In our shops we work nothing but the best Swedish we can buy which has cost us lately 19/6 p^r cwt. If you c^d hear of any Spanish Iron cheap I will send proper sizes & I think we c^d apply it usefully. As far as human foresight can extend we sh^d not expose ourselves to such accidents if can be prevented. I have totaly forbid the making any more Iron work at Bersham. I have kept all the specimens of this Gudgeon for Mr W^s inspection & to stay his stomach in the meantime have treated him with a letter on that & on his Boiler plates but have yet said nothing of the accident which I believe broke the Gudgeon, as he is too ready to catch hold of any excuse.

A part of the engine that gave trouble was the beam. At first these beams were plain oak logs, then Smeaton introduced beams built up of several logs, but with the advent of the double-acting engine the repeated reversal of stress to which the beam was subjected made it peculiarly liable to failure. Watt had adopted the beam trussed with a king-post in 1778 (see Fig. 6, p. 98); he used also beams with diagonals from the gudgeon tied together and to the ends of the beams. Now in the Science Museum are preserved two models of such beams; in one case hooks and in the other bridles (see Pl. XI (b)) are provided on which to hang weights, obviously for testing them. Beams of this type were used in a number of engines from 1786 to 1798, shortly after which the cast-iron beam came into use. Also in the Science Museum there are two models of framed structures of the Warren girder type, obviously meant for beams; one of these has been tested to destruction. Beams of this type were not used in practice, however. Such was Watt's pioneer work in the testing of materials.

With the introduction of the rotative engine, Watt was faced with the question of how to charge for them. In the case of the pumping engine the charge was based, as we have seen, on the savings effected in fuel consumption as compared with an engine that the new one replaced. When the rotative engine was introduced an analogous situation arose; frequently an engine was wanted to replace the labour of animals, hence the number of horses that the engine could replace naturally became a measure of its performance. An annual premium based on the horses' power which the engine was capable of exerting was the obvious way of charging for it.

The idea of comparing the power of an engine with that of horses was no new one. Thomas Savery had pointed out in 1702 that one of his engines would raise as much water as two horses working at the same time, i.e. ten or twelve relaying each other in shifts if the work were continuous. Smeaton in 1775 gave a striking illustration of the power of his atmospheric engine at Cronstadt in Russia, when he said it was "equal to the labour of 400 horses", but no idea of a unit of measurement was in his mind. At the outset Watt, too, was concerned only with systematising his work as an engine builder.

Millwrights concerned with horse gins knew pretty well what a horse could do and we find Watt in the "Blotting and Calculation Book" already mentioned, in 1782, based his calculations on data supplied to him to the effect that a mill horse walks, in a path of 24 ft. diameter, $2\frac{1}{2}$ turns in a minute. Watt assumed that the mill horse exerted a pull of 180 lb.—we do not know where he got this figure—and found that it exerts 32,400 lb. per minute. By the following year he has rounded off the figure to 33,000, doubtless for ease in calculation. Then in 1783, we get the complete statement: "Each horse = 33,000 lb. 1 foot high p. minute." Within a few years it was the firm's ordinary practice to rate their rotative engines at so many horses, i.e. a "20 horse engine". Little did Watt foresee that he was setting up a unit that would come into use throughout the world. The unit is defined in print correctly for the first time in the *Edinburgh Review* for 1809, in the course of a review of Gregory's *Mechanics*, wherein that author said: "What is called a horse's power is of so fluctuating and indefinite a nature that it is perfectly ridiculous to assume it as a common measure by which the force of steam engines and other machines should be appreciated." The reviewer took Dr Gregory to task on this loose statement and supplied the correct definition given above. It may be remarked that for long afterwards, engineers generally did not recognise that the term "horse-power" connotes a rate of doing work, expressed in pounds, feet and minutes.

By the end of 1785 Boulton had paid off his overdrafts

and advances from the bankers and was in like happy case as Watt. But how different the attitude of the two men! The latter would have been content to have retired with his modest competence, but Boulton only thirsted for "fresh woods and pastures new". We cannot pause to refer to his activities outside the engine business, but within that field his conviction was that the number of applications of the rotative engine in industry was enormous. One such application in which he took particular interest was that to flour milling, hitherto performed to the satisfaction of millers and proprietors by wind and water power. Among the Watt models in the Science Museum are two showing the application of the engine to this duty, and there is little doubt that they were schemes of Boulton.

An opportunity occurred to put his ideas into practice —a project for a steam flour mill in London was put forward in 1782 or 1783. It was not the first of its kind, for Wasborough had erected one at Bristol driven by a common engine, but the Albion Mill, for so it was called, situated on the Surrey side of the Thames near Blackfriars Bridge, was a much more ambitious scheme. Boulton entered into it with spirit, backing it with capital and inducing others, Watt among them, to do likewise. Buildings, from the plans of the well-known architect, Samuel Wyatt, were erected, with provision for three sets of engines and millwork. The construction of this machinery was undertaken by Boulton and Watt. Engine No. 1 was of 50 horse-power, cylinder 34 in. diameter by 8 ft. stroke, driving 10 pairs of millstones

$4\frac{1}{2}$ ft., of which only six were in operation at one time. The output of each pair of stones was 10 bushels per hour. Engine No. 2 only differed from No. 1 in detail. The third engine was not installed for reasons that will presently appear. A number of new ideas were embodied in the machinery: cast iron was used exclusively in the gearing in place of timber. No. 1 engine was one of the first to embody Watt's parallel motion—it was in fact designed with the three-bar motion that has been already mentioned. Then engine power was applied to all purposes at the mill—unloading wheat from barges, sack-hoisting, sifting and dressing the flour. The design and construction of the machinery for the latter two duties was entrusted, on the recommendation of Boulton and Watt, to John Rennie, the Scottish millwright; it was his first introduction to London and it was largely on the reputation established by his work on the mill that he became so well known subsequently as a civil engineer. He was engaged on the milling department from 1784 to 1788, while Boulton and Watt were putting in the engines. The Albion Mill came to be regarded as the great mechanical wonder of the day.

As a commercial enterprise the mill bid fair to be a great success and bring in a large return to the proprietors, but on March 3rd, 1791, barely three years after its completion, it was burned down. It was believed that this was the result of incendiarism and not an accident, but this could never be proved. What lent colour to the belief was that the undertaking had been viewed with great hostility by the millers, and in fact the proprietors had been calumniated grossly on the ground that they were establishing a monopoly. Of course the public is always ready to believe such a charge if repeated often enough, but that it was entirely unfounded is shown by the fact that the mill was the means of lowering the price of flour in the metropolis. From their limited and selfish point of view the millers were right; indeed they may have had the inkling that this mill marked not only the beginning of the transfer of flour milling to seaports, there to be practised on the large scale, but also the doom of the country wind or water mill serving local needs on a small scale.

The Albion Mill had been badly organised and managed and it was not rebuilt. Boulton lost, it is said, £6000 and Watt half that sum, severe blows to both of them. With this brief account of the mill in mind, the two letters below from Watt are quite interesting. The earlier of the two (1785, Nov. 5) is evidence of some of the false reports about the mill that had been set in circulation and had come to Boulton's ears when in town. The later letter (1786, April 17) shows Watt's intense aversion to advertisement and incidentally proves what we have said above about bad management.

As to the paragraph you mention, I suppose it was put in by some wiseacre or more probably by the millers, as the building has never reeled a iota, but as it has no walls & consists merely of floor posts set upon one another, the props you mention were put up on *all* sides to *prevent* reeling. All the work at the mill seems extremely well

done, the only thing wrong seen hitherto is expence
which I fear will be very great.

It has given me the *utmost* pain to hear of the many
persons who have been admitted into the Albion Mill
merely as an Object of Curiosity— Were there no other
loss than the taking up your time it is a very serious one
but there are other essential ones which are too obvious
to need to be pointed out, among which are that the
disgraceful condition in which it has hitherto been has
been more likely to do us hurt than good as engineers &
the bad management or want of management in other
respects must hurt the credit of the Company— I hear
from different quarters enough to convince me that we
are looked upon by the serious common sense man as
vain and rash *adventurers* that our talking of what we can
do is construed into either a want of ability to perform it
or the foolish cry of Roast beef [i.e. the announcement of
one's good fortune]—my natural hatred of ostentation
may perhaps make me feel these things too strongly, but
surely those who say so think they have some reason for
the observations & it cannot happen that the most
pointed of them can come to my ears, considering how
little company I keep— Among other things I heard
some time ago that on a certain day there was to be a
Masquerade at the A M, and this from persons no ways
connected with us & who had heard it as com^n Birm^m
talk—and I felt it as a severe reproach considering that
we are much envied at any rate, everything which con-
tributes to render us conspicuous should be avoided, let
us be content with *doing*. R. [i.e. John Rennie] no doubt
has vanity to indulge as well as us but he sh^d be curbed
& the bad consequences pointed out to him, it will ruin
him, Dukes & Lords & noble peers will not be his best
customers— And let me entreat that the doors of the

Mill be strictly shut against all comers without an order signed by three Comittee men & that only at a comitte meeting on some fixt day of the week & let that rule be inflexibly adhered to. I know that you have been actuated by good motives in showing the mill, namely the desire of getting quit of part of the property we have in it & the hopes of making interest to get a charter, but I conceive these things will be better attained by making it a mystery to the many & by the external appearance of business.

I cannot think on corn machines till every thing else is successful— In the meantime the corn sh^d be frequently turned & the flour attended to, for it will suffer much more than the corn.

In spite of the fact that Watt now had help in the drawing office, he still, like so many men, tried to do too much himself, and when mistakes occurred, as they invariably do, he got very angry. The following letter is typical of many more; the last paragraph especially is quite in his best pessimistic vein. His dictum that "mistakes may be fallen into by any body" is one that will find an echo in everyone's breast. The letter of Watt to Boulton, who was at Chacewater, is dated 1785, Nov. 5:

I am exceedingly vexed at the omission of marking the distance of the centre of the perpendicular radius of Wheal Fortune below the horisontal line of the centre of Gudgeon, it should have been markt 42 inches, the perpendicular link being ⅛ inch shortned by the angle it makes.

It is drawn right in the General section which I wish had been consulted, and indeed in new subjects of that kind which are so complicated, it should be a rule to make some kind of rough model to see that it is right as mistakes may be fallen into by any body....

On the whole I find it now full time to cease attempting to invent new things, or to attempt anything which is attended with any risk of not succeeding, or of creating trouble in the execution. Let us go on executing the things we understand and leave the rest to younger men, who have neither money nor character to lose.

About 1786 the partners meditated the acquisition of privileges or patents for the rotative engine in other countries. The State of Connecticut and the States of Holland may be mentioned, but more important was an invitation from the Government of France of that day to visit Paris in order to estimate for an engine to replace the celebrated, but noisy, and hopelessly inefficient machine of Marly, that supplied the palace of Versailles with water from the Seine. However, the fall of the minister, M. de Calonne, and the oncoming of the French Revolution, brought negotiations to a standstill. The visit to Paris was Watt's first introduction to the Continent and widened his horizon by contact with such scientific minds as Lavoisier, Laplace, Monge, De Prony, and Berthollet. The latter had discovered in 1785 the method of bleaching by chlorine gas and demonstrated it to Watt. In 1787, the latter communicated the process to his father-in-law, James MacGregor, who tried it out under Watt's direction at his bleach-works near Glasgow. The intolerable stench and danger to health prevented its adoption, and it was not until Sir Charles Tennant discovered the way of making bleaching powder by the action of chlorine on slaked lime that the new method superseded the old and tedious one of bleaching by sunlight.

While Watt was in Glasgow he received from the firm's bankers in London the pleasant news that £4000 had been transferred to his private account. Watt invested it carefully, whereas Boulton's only use for capital was, as we have said, to launch out afresh. This time it was to form a kind of trust to steady the market in copper. When in the following year, owing to the commercial crisis that occurred then, he was unable to hold on, he wrote (1788, May 4) to Wilson: "Mr Watt hath lately remitted *all* his money to Scotland and I have lately purchased a considerable quantity of copper. . . . I shall be in a very few weeks in great want of money and it is impossible to borrow in London or this neighbourhood as all confidence is fled." From this one would infer that Boulton had tried to borrow from Watt but that the latter had excused himself from lending on the ground that he had no ready money handy. One would have thought that Watt would have done everything he could to relieve his partner's predicament and one can only call Watt's conduct shabby.

We have, however, somewhat outrun the development that was occurring in the rotative engine, the sale of which was exceeding even Boulton's sanguine expectation. The engines were erected in a similar fashion to the pumping engines, but were charged for, as we have said, since the payment based on savings in fuel was inapplicable, on a fixed premium based

PLATE XII. THE "LAP" ENGINE, 1788

Courtesy of the Science Museum

on the horse-power, i.e., the premium was six guineas per annum in the London area and £5 in the provinces.

A typical factory engine of this period and construction has fortunately been preserved and is to be seen at the Science Museum (see Pl. XII). It is the engine built for Boulton in 1788 for driving the laps or polishing buffs in Soho Manufactory itself; hence it was known familiarly as the "Lap" engine. A point of exceptional interest about this engine is that it was the first to which Watt applied his centrifugal governor.

Although Watt may not have been the first to suggest the application of a governor to a steam-engine to regulate its motion automatically, there is no doubt that he was the first to apply it. Some form of centrifugal governor was in use at the Albion Mill, where Boulton saw it and was so much struck by it that he described it in a letter to Watt (1788, May 28) as being for regulating the pressure or distance of the top mill stone from the bed stone in such a manner that the faster the engine goes the lower or closer it grinds and when the engine stops, the top stone rises up. . .; this is produced by the centrifugal force of 2 lead weights which rise up horizontal when in motion and fall down when ye motion is decreased by which means they act on a lever that is divided as 30 to 1 but to explain it requires a drawing.

This was all that was necessary for Watt's acute mind. Five months later, in the firm's "Drawings Day Book", recording work done in the drawing office, we have an entry under November 8th, of a "Drawing of centrifugal engine regulator for no. of strokes". On December

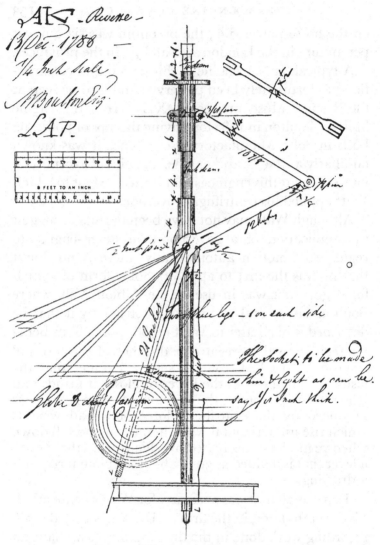

Fig. 10. Watt's centrifugal governor

11th, Southern started a new drawing which he finished and dated "13 Dec. 1788"; it is marked "M. Boulton Esq. LAP", which means that it was intended for that engine. Our illustration (Fig. 10) shows this notable drawing. It will be seen that heavy balls, two in number, are attached by link-work to a vertical shaft so that, when the speed increases, they fly outwards under centrifugal action, and in doing so raise a sleeve which by a lever (not shown) actuates a butterfly or throttle valve in the steam pipe leading to the engine. The supply of steam is thereby diminished and the speed of the engine consequently reduced till equilibrium is again established. Watt did not patent the governor; maybe he felt that he was on unsafe ground, for the mechanism already mentioned as in use in flour mills might have been cited against him as an anticipation.

A few words may usefully be said about boilers because the impression is widespread that Watt made improvements in them; this, however, is only partially true. The boiler that supplied steam for the common engine was in principle nothing more than a brewer's copper. In appearance it resembled a haystack—circular in section, dome-shaped top, tapering sides and flat or convex bottom; it went by other names such as the "round", "beehive" and "balloon" boiler. It was set in brickwork with flues arranged to cause a wheel-draught. Since the boiler worked at atmospheric pressure, or thereabouts, it was little more than a tank, tight enough to keep steam from escaping into the air and *vice versa*—and a tank it remained to the end of

Watt's time. Watt supplied boilers as called for by local requirements. In Cornwall a long boiler of the same cross-section as the haystack, known as the waggon boiler, was preferred, and it is this one that is generally attributed to Watt, although all that he claimed for himself was that he "somewhat improved the form and adjusted the proportions"—we cannot imagine him doing less! A typical boiler setting of the kind is to be seen in Fig. 11, p. 172. Even with regard to boiler furnaces Watt says: "The conveying of flame through flues in the inside of the water had been practised by others before my time."

Although Watt made no fundamental improvements in boilers, he did concern himself with the prevention of smoke from boiler firing. Writing to De Luc (1785, Sept. 10) Watt says: "I have some hopes of being able to get quit of the abominable smoke which attends fire engines. Some experiments I have made promise success. It is not on Mr Argand's principle, but an old one of my own, which is exceedingly different." By that time he had already obtained a patent (June 14th); it was for a furnace "whereby greater effects are produced from the fuel, and the smoke is in a great measure prevented or consumed".

Describing the experiments to his wife (1785, Oct. 9) Watt says: "We had a first trial yesterday of a large furnace to burn without smoke under the big boiler at Soho that used to poison Mr B's garden so much; and it answered very well as far as we could judge from a wet furnace, and without the engines being at work."

It is quite possible that the ill-effects of smoke from the boilers on the vegetation in Boulton's garden—the grounds of Soho House adjoined the Manufactory—was the prime cause of Watt giving his mind to the problem of smoke prevention. The scheme did not succeed altogether, but what we can say is that Watt was the first to bring out an apparatus to combat the anti-social evil of smoke emission from factory chimneys.

We have mentioned already at some length the Lunar Society, its membership and proceedings, but a further word is necessary owing to the entry into the charmed circle in 1780 of Dr Joseph Priestley; his personality and researches greatly impressed the members, among whom at once a rage for chemistry set in. It was Priestley's experiments that led Watt in 1782 to make the brilliant deduction that water was a compound substance and not an element. The discovery was claimed also for Cavendish and a heated controversy raged between their respective partisans as to who was entitled to the credit. It is unnecessary to go into the merits of the case since neither Watt nor Cavendish enunciated the doctrine of the true nature of water held to-day; that was reserved for Lavoisier.

We must give one instance of the "Proceedings" of the Society. Mr Kirwan had sent in a communication on the subject of Scheele's discovery of prussian blue. This communication was read to the Society on November 10th, 1783, by Watt who, four days later, wrote to Kirwan thanking him in the name of the Society for his paper. The interesting part of this letter is what follows:

Having lately been making some calculations from Messrs. Lavoisier and De la Place's experiments and comparing them with yours, I had a great deal of trouble in reducing the weights and measures to speak the same language....It is therefore a very desirable thing to get all philosophers to use pounds divided in the same manner....My proposal is briefly this; let the

Philosophical pound	consist of	10 ounces	or	10,000 grams
the ounce	,,	,, 10 drachms	,,	1,000 ,,
the drachm	,,	,, 100 grains		100 ,,

Let all elastic fluids be measured by the ounce measure of water; by which the valuation of different cubic inches will be avoided and the common decimal tables of specific gravities will immediately give the weight of those elastic fluids....I have some hopes that the foot may be fixed by the pendulum and a measure of water, and a pound derived from that.

This is, it will be seen at a glance, nothing less than a decimal system of weights and measures based on an absolute measure. In January 1784, having given the matter further consideration from the practical aspect, Watt gave his "preference to those plans which retain the foot and ounce". If nothing else came of these proposals, they added to the volume of opinion in favour of reform of standards of weights and measures. It is too much to claim, however, as some have unwisely done, that Watt's suggestions were the basis of the metric system introduced in France after the Revolution.

With what regard the Lunar Society was held by its Members is best told in their own words by those whose lot it was to be deprived of the intercourse it afforded. Erasmus Darwin, when he removed from Lichfield to Derby in 1782, wrote this sparkling note to Boulton: "I am here cut off from the milk of science which flows in such abundant streams from your learned Lunatics, and which, I assure you, is a very great regret to me." Again, when Dr Priestley was obliged to remove from Birmingham owing to the Church and King Riots, he wrote in similar strain to Watt of "the pleasing intercourse I have had with you, and all my friends of the Lunar Society. Such another I can never expect to see. Indeed London cannot furnish it. I shall always think of you at the usual time of your meeting." With this gracious farewell, we too may fittingly close our account of the Society.

We have now arrived at the close of the brilliant period of Watt's life, when he began to think that he deserved release from business and that the time had arrived for him to take life more easily.

CHAPTER VII

PARTNERSHIP WITH MATTHEW BOULTON. CLOSING YEARS, 1790–1800

Heathfield. Education of sons to become junior partners. Soho Foundry. Infringers and pirates. Litigation. The patent triumphant. Doldowlod. Termination of partnership.

IT is a sure sign that a man is prospering when he decides to remove to a better house than the one he is living in. In the case of Mr and Mrs Watt this was not the whole story, because at the time of which we speak Harper's Hill was being invaded by the tide of bricks and mortar from Birmingham, and Regent's Place was becoming less and less desirable as a place of residence. In 1790 Watt acquired land on Handsworth Heath and began under the direction of the architect, Samuel Wyatt, already mentioned, the erection of a roomy mansion, appropriately named Heathfield. The situation was pleasant and the distance from Soho only some twenty minutes' walk. The enclosure of the Heath in 1791 afforded Watt an opportunity of buying more land and in this way he obtained possession eventually of about 40 acres. He proceeded to plant timber, wall in a garden, erect stables and outhouses and build a couple of lodges. The house was well appointed; some cupboards in the dining room gave the impression that Watt himself had had a hand in their design.

PLATE XIII. WATT'S HOME AT HEATHFIELD, STAFFORDSHIRE, c. 1895

Courtesy of J . H. Tangye, Esq.

Heathfield grew into a spot of silvan beauty—our illustration (Pl. XIII) shows what it looked like about 1895—but the advance of the builder has proved insistent and we regret to say that in 1927 the house was pulled down and the last of the estate sold.

Watt was gradually easing off from business and so to a slight extent was Boulton, for each of them was training a son to follow him in business and was looking forward expectantly to seeing his son take up the position of a junior partner in the firm.

Boulton had an only son, Matthew Robinson (his second Christian name was his mother's maiden name), born in 1770, who was brought up with great care. Boulton sent him to Paris to acquire a knowledge of French, and while there much kindly advice passed by letter from father to son—he was implored to keep out of bad company. The father's great hope was to see his character develop: "There is nothing on earth I so much wish for as to make you a *man*,—a good man, a useful man, and consequently a happy man." It will occasion no surprise to learn that the young man spent more than his father thought was good for him. We have an interesting glimpse of him by a lady who saw him when home from Paris in the summer of 1788; Boulton introduced him to a meeting of the Lunar Society, and she says: "I well remember my astonishment at his full dress in the highest adornment of Parisian fashion." She observed that the members "with one accord gathered round him, and asked innumerable questions . . . the party heard, no doubt, in this young man's narrative, the distant, though as yet faint, rising of the storm which, a year later, was to burst upon France".

Young Boulton went back to Paris to finish his studies and developed to his father's satisfaction into a man of ability and character. He returned home on the eve of the French Revolution. All thinking minds were affected by the event and none more so perhaps than Dr Priestley, who forgot for the time his scientific studies and wrote and preached of the brotherhood of man and of the downfall of priestcraft and tyranny. He welcomed with enthusiasm the acts of the National Assembly of France, abolishing every form of feudal institution. By doing so, he brought himself into public prominence and exasperated local feeling, an occasion for exhibiting which soon occurred. About eighty local gentlemen gathered at the "Hen & Chickens", then the best inn in Birmingham, on July 14th, 1791, to celebrate the second anniversary of the French Revolution. A mob assembled, broke the inn windows and with shouts of "Church and King" went off to the chapel where Priestley ministered. Having set fire to and gutted the place, it being now nightfall, the mob made for Dr Priestley's house at Fairhill. He had not been present at the dinner but nevertheless had had warning of the mob's coming and had escaped with his family only half-an-hour before their arrival. The place was sacked and the records of twenty years' scientific work burnt with the rest of the contents of the house, as was the house itself. The "Lunatics" were marked down especially for attack. Boulton and Watt barricaded Soho and armed their workmen, but fortunately the rioters did not come that way. Some of the arms served out on this occasion are still to be seen in the Boulton and Watt Collection.

Among those who were carried away with enthusiasm for the French Revolution was young Watt, but before giving an account of the active part he took in it, we must state, in brief, what had been his upbringing. At the age of fifteen his father sent him to John Wilkinson's Bersham Ironworks in Wales where, so his father told a friend (1784, May 30): "He is to study practical book-keeping, geometry and algebra in his leisure hours; and three hours in the day he works in the carpenter's shop." We have no mention of any tutor being provided, but the allocation of employment not only shows a belief in the value of craftsmanship but suggests the germ of the sandwich system of engineering education. After remaining a year at Bersham, as his father had intended, the youth was sent to Geneva where he studied general subjects, but especially languages to such effect that he became fluent in French and German.

On his return to England in 1788, Boulton found a place for him in the counting house of Messrs Taylor & Maxwell, fustian makers, of Manchester, where he remained two years. Boulton's and his father's reputation served to introduce him to a large circle of friends. Then occurred what seems almost inevitable—the allowance from his father proved insufficient. Did he appeal to him for help? Nothing of the kind. As we have seen already, Watt was undemonstrative and averse to showing feeling; in this respect his attitude towards his own family seems to have been in no way different to that towards outsiders. Young Watt wrote to the generous, the sympathetic Boulton, a well-expressed, manly and respectful letter asking for a loan of £50.

Receiving no reply from Boulton and fearing greatly that the letter might have got into wrong hands, besides becoming anxious about his creditors, young Watt wrote again to Boulton. The delay in replying on the latter's part had been due to absence in London and on Boxing Day, very appropriately, he sent the young man a draft for £50, the sum asked for, together with fatherly counsel, to which the young man promised in a very proper letter (dated 1789, Dec. 30) that he would conform.

Among the friends made by young Watt in Manchester was Thomas Cooper, M.D., then engaged in an extensive bleaching business, who was keenly interested in politics and particularly in the thrilling events then taking place in France. He embued Watt with his principles, and as members of the Constitutional Society of Manchester these two were appointed delegates to present an address to the Société des Amis de la Constitution. This they did in Paris on April 13th, 1792, an action which was denounced by Edmund Burke in the House of Commons a week later. Young Watt is said to have been on intimate terms with the Jacobin leaders. Both Cooper and Watt fell under the suspicion of Robespierre, who denounced them as secret emissaries of Pitt. It is said that Watt sprang into the tribune and defended himself with such eloquence that he carried the Assembly with him. From that moment, however, he felt his head insecure, so fled to Italy whence he went on to Germany and did not return to England till 1794.

Watt was under great apprehension for his son's safety, for many members of political societies had been arrested and lodged in the Tower. However, young Watt said "he never corresponded with any of them [i.e. the Societies] at any time... that for these two years he has had no sort of a connection with any of them, and for more than a year all his correspondence has been recommending his friends not to intermeddle with public affairs". Evidently this is the decision he had come to for himself. The storm blew over and subsequently to October 1794 when he was admitted a partner in the firm he devoted himself wholly to business and never took part in public affairs. The author has always surmised that there must be some deep reason for this *volte face* on the part of young Watt, possibly a love affair; it is significant that he never married.

It must have been a great relief to their parents when the young men thus settled down and entered the business, for they were travelled, linguists, trained in business and of higher attainments generally than persons who were carrying on engineering works elsewhere. It was the determination of the two elder men that the sons should be taken into partnership and this was shortly after carried into effect. First, however, the sons tried their wings as it were, by taking over between them in 1794 the copying-press business, the elder men retiring in their favour.

In October 1794 a new firm under the style of Boulton, Watt & Sons was formed, the partners were Boulton, his son, Watt and his sons James and Gregory; the latter was a boy of nineteen, still at college, of whom more anon.

It was in fact the parting of the ways in another sense. The old firm had obtained its profits from the premiums on engines. With the lapse of the original patent, now imminent, these would cease and unless the business and goodwill were to pass entirely out of their hands it was imperative that the policy of the firm should be altered entirely, and that they should become engine makers themselves in competition with the rest of the world. It was probably Boulton who saw this most clearly, for was it not he who in 1769 had foreshadowed (see p. 52) the production of engines in a works specially built and laid out for the purpose? The bold policy of erecting an entirely new establishment with every facility that experience could suggest for the manufacture of machinery was decided upon. The idea was put into concrete form and Soho Foundry was inaugurated. What a splendid outlet for the energies of the young partners! Land was purchased in Smethwick in 1795, conveniently situated for transport by the side of the Birmingham and Wolverhampton Canal, at a spot only about a mile distant from the Manufactory. The buildings erected comprised forge, smithy, boring mill, turning, fitting and carpenters' shops, drying kiln, foundry and air furnace. A wet dock from the canal was excavated. The money for building and equipping the new works was advanced by Boulton and by Watt jointly. Peter Ewart, a millwright, who had previously been employed by Boulton in the erection of millwork and machinery for his mint, and by the firm for the erection of engines, was engaged to help in the work of construction and lay-out of the machinery. It has to be realised that there were then no machine tools beyond the lathe and the boring mill and further that there were no machine tool makers. Each firm made their plant for themselves and each workman had his individual kit of tools. In this case, the boring mill was the most important tool of all and on it Ewart exerted his skill. It embodied the central bar, the important feature of Wilkinson's mill, but it was vertical instead of horizontal. It was not at first a great success, but after the departure of Ewart, about 1798, Murdock, who was recalled from Cornwall about that time, completed the mill.

There is no evidence that Watt took any great share in establishing the new Foundry; he was content with advising generally. The staffing of the foundry was a matter of difficulty as trained men were hardly to be had. However, men from Wilkinson's works—men who had worked on the Boulton and Watt engine parts—were engaged and gradually the place was got into working order.

The Foundry proved to be an exceedingly wise and profitable venture. Castings and forgings could be obtained when wanted and the delays of the former system obviated. The work grew and the difficulty of keeping pace with it necessitated extensions. The standard of workmanship at first was not so high as that of other firms, such as Matthew Murray of Leeds, but great pains were taken by the junior partners not only to emulate those firms but to scour the country for good men, especially moulders, to improve the technique of the rest. As time went on Soho Foundry became a school of practical engineering and to have been apprenticed or employed there was a recommendation for any engineering post. The Foundry was a very paying concern: by September 1804 the total capital cost amounting to £27,431 was paid for out of profits and by the end of 1812 the whole of the debt was cleared off. Not only so, but by 1816 further extensions costing about £20,000 were likewise paid for out of profits. The truth is that the really palmy days, from the money-making point of view, of the firm, now Boulton, Watt & Co., supervened after the original partners had retired.

A distinguishing feature of the Foundry was the way in which it was organised and run. There is nothing in the costing, the feeds and speeds, the scientific management of Taylor, Ford and other experts of the present day that was not to be found at Soho before 1805. This early engineering works possessed an organisation that far excelled that of any other extant establishment and that can bear comparison with that of any works of the present day.

Watt seems to have given close attention to the improvement of engine performance. We quote from a letter, written to Southern (1792, Sept. 22) from Truro, which is typical of dozens of others. The letter is as follows:

Since we wrote you last have finished our trial at Wheal Butson as follows

36 inch Cyl^r = beam exactly 8 feet stroke
pump 9½ inch bore, Depth 50 fath. 2 feet = 9291 lb

First 12 hours burnt 22 bushels made 7560 strokes/ strokes p^r bushel 343·6 effect 25,539,100 lb to one foot high p^r bushel. N.B. The fire was out of course when begun and 12 bushels of the 22 were thrown in in the first hour.   strokes

Second 12 hours burnt 17 bushels of coals, made 7550 effect 444 strokes p^r bushel = 33,001,632^{lb} p^r bushel to 1 foot

The regular effect of Poldice double 58 in the month of Augt is equal to 32 million, burns only 4½ bushel (100 p^r day) p^r hour for 6 strokes p^r minute Column water = 63468^{lb} stroke 9 cyl^r 6 pumps

We have been at Tin Croft the account of which for last month from the Captains is

fath						bore				lb
17	3	9	...	...	...	9·3	...	...	...	3117
10	3	3	...	...	...	8·5	...	...	...	1551
4	1	0	...	...	...	8·25	...	...	...	589
						Total load	...			5257

Coals 22 bushels p^r day strokes 7½ of 5 feet 9 inches long p^r minute = 491 p^r bushel, effect 14,841,825^{lb} to one foot high— No body or few will own conviction though we have circulated accounts of these experiments. But we have reason to believe they have taken considerable effect upon mens minds.

The data given in the letter on the performance of the engines show that the "duty" (already defined previously) had reached the relatively high figure of thirty-three millions. When it is remembered that the duty of the common engine was only about seven millions, and that even with Smeaton's improvements the duty only attained ten millions, we can realise what an enormous saving Watt had effected. Further, when we mention that he lived to see the engine attain, although in other hands than his own, a duty of about fifty millions, the comparison is still more striking.

We have seen above how Watt determined exactly what duty a pumping engine could do, but as to the performance of his mill engines he was on anything but sure ground. The engines were rated at so many "horses", but there was no direct means of finding out whether they actually exerted the rated power. As a step in this direction, somewhere about 1790, Watt invented an instrument which he called an indicator, but which we should call a pressure gauge, whereby the variations of pressure during one stroke of the engine could be observed. The instrument was coupled up by a pipe to the engine cylinder; it had a cylinder with a piston which was pressed on by a spiral spring and the rod of the piston carried a pointer moving against a fixed scale. In a modification of the instrument, brought out a few years later, the piston rod worked against a spiral spring through the intermediary of a beam and on the latter was mounted a pointer passing over a fixed scale. One of these instruments (see Pl. XIV) is preserved in the Science Museum.

It was not a big step, but it was a most brilliant one, to substitute in the first-named apparatus a pencil for the

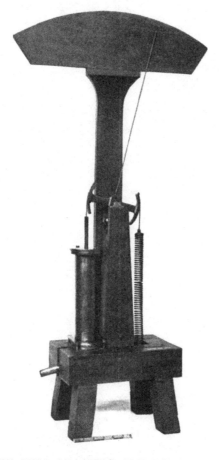

PLATE XIV. WATT'S INDICATOR OR
PRESSURE GAUGE, *c.* 1794
Courtesy of the Science Museum

pointer, and to furnish a board or tablet moved by the engine and carrying a sheet of paper on which the pencil could trace its movements. This gave a diagram which showed everything that happened in the cylinder; the area of the diagram too was a measure of the work done. This invention was originated in 1796 and was almost certainly due to Southern. It still continued to be known as the indicator, however, and its successors have continued to be so known to the present day although it is a misnomer; work-measurer or ergmeter would be a more appropriate name.

The patent for the crank as already mentioned expired in 1794. The firm naturally adopted the crank and we give a representation of one of the first crank engines, made for the firm of William Hawks & Co. in 1795 (see Fig. 11). The state of development of the rotative engine at that date is well shown. The framing is still entirely of timber as is likewise the working beam and the connecting rod. The cylinder 19 in. diam. by 64 in. stroke is carried on a timber frame while the beams that support the working beam are built into the engine house walls. It should be remarked that the boiler steam pipe is marked "to be very well wrapped", an early instance of pipe covering or heat insulation now universally employed in such situations.

By far the greatest amount of Watt's time during the decade under review was taken up by litigation. At the time described in a letter (1792, April 21) to Southern, he was in London attending at the House of Commons, opposing the Bill introduced by Hornblower for ex-

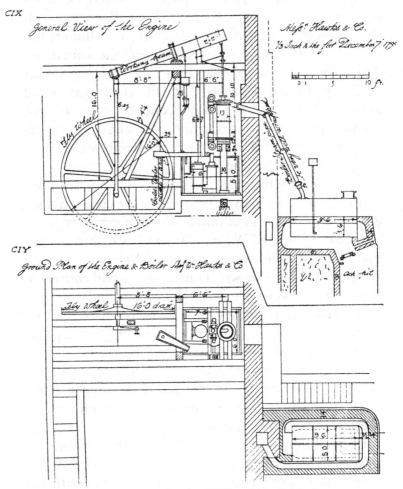

Fig. 11. Rotative engine for William Hawks & Co., 1795

Boulton and Watt Collection. Courtesy of Birmingham
Reference Library

tension of his compound engine patent already mentioned: "As they have brought Mr Giddy the high sheriff of Cornwall, an Oxford boy, to prove by *fluxions* the superiority of their Engine, perhaps we shall be obliged to call upon you to come up by Thursday to face his fluxions by common sense." It was a compliment to Southern to tell him that he was to be called as a witness to face Giddy, because the latter was the well-known Davies Giddy, who subsequently changed his name to Gilbert and became President of the Royal Society.

The same litigation is referred to in the succeeding letter (1792, Sept. 15), also to Southern: "The people here are in general as absurd as ever nor do I believe it possible for an angel to convince them of the inferiority of Hornblower's engine though it is now nearly unable to keep the bottoms dry."

As we have stated already, Hornblower failed to secure an extension of his patent and thus that bogey was laid to rest. The real tug-of-war was to come, however. There had been a succession of pirates, schemers and infringers of the patent, but either they gave in without the necessity of taking legal action or else they were too unimportant to be worth powder and shot. Out of the whole crowd, however, there were two who persisted in their opposition to the firm until the very last. These were Jabez Hornblower, brother of the Jonathan who had patented the compound engine, and Edward Bull.

Before dealing with these opponents we may perhaps offer one or two general observations. One hesitates to say so, but it is a question whether Parliament in its wisdom did not err on the side of granting too long an extension to Watt's patent. The enormous increase in coal-mining, iron manufacture and industry generally in the period under review could perhaps not have been foreseen, but the demands were wholly beyond the capacity of a single firm to supply, hence if one did not infringe, and could not wait possibly an unreasonable length of time for a Watt engine, one had to be satisfied with an inefficient or inferior engine. Watt, as we have said, resolutely set his face against granting licences to use the separate condenser—a course that might have eased matters. Then the premiums exacted by Boulton and Watt were no slight burden upon industry. Again, those who were paying premiums were witnesses of the spectacle of pirates and infringers erecting engines and refusing or avoiding payment of premiums.

Boulton and Watt were reluctant to take legal action for a good reason: Watt, bearing in mind the experiences of other inventors and aware of the defects of his own specification, was not too confident that the patent would be upheld if taken into the Law Courts. Having deferred the evil day as long as they could Boulton and Watt were forced eventually to take action.

Jabez Hornblower, a member of the well-known and large family of Cornish engineers of that name, was employed by the firm in 1779 in Cornwall as an erector; he was of a bad-tempered and ungracious disposition, however, and the firm did not continue to engage him. After a varied experience, he gravitated about 1790 to

London and there set up as an engine maker. In 1795 he entered into partnership with one Maberley who had acquired the patent taken out in 1791 by Isaac Man-waring for a double cylinder engine. Maberley improved it by connecting the piston rods of the two single-acting cylinders by a chain lapping over a wheel above, thus in effect making it into a double-acting engine. So far this was quite in order and no exception could have been taken; the engine, however, embodied the separate condenser and this constituted an infringement of Watt's patent. Injunctions were served on the parties by Boulton and Watt. As Watt wrote (1796, March 20): "The rascals seem to have been going on as if the patents were their own.... We have tried every lenient means with them in vain and since the fear of God has no effect upon them, we must try what the fear of the devil can do." The case came on in the Court of Common Pleas on December 16th, 1796, before Lord Chief Justice Eyre and a special jury. The verdict was for the plaintiffs. When the result became known at Soho there were great rejoicings, firing of cannon and what not. The defendants, however, brought proceedings in the same Court on a writ of error, on the ground that the patent itself was invalid. In the main the objections were to the sufficiency of the specification, alleging that the description was not enough to enable the public to construct the engine—the condition of course on which the patent was granted originally. Eventually the judgment was affirmed but this did not take place till 1799 because the other action for infringement, Boulton and Watt v. Bull, to which we must now refer, turned eventually upon the same legal point.

Edward Bull had been employed by the firm as an erector in the Midlands and in 1781 had been sent into Cornwall, where as we have seen engines were being erected apace. Naturally he became well known among the mine adventurers, who found him pliable and easily withdrawn from the firm's interests. In 1792 he brought out the engine that goes by his name, the chief feature of which is the inverted cylinder with its piston rod directly coupled to the pump rod. Now Watt had schemed such an engine in 1766—our illustration, Pl. VI, p. 58, shows this—and two years later he had actually made one but had not followed it up. This, however, was a matter of unimportance; the real point was that Bull's engine embodied the separate condenser. An injunction was served upon him and, as the firm was under the impression, although wrongly as it turned out later, that his assistant Richard Trevithick was in partnership with him, the firm issued an injunction against Trevithick also.

The cause Boulton and Watt v. Bull came on for hearing in the Court of Common Pleas on June 22nd, 1793. Watt was in town in attendance, and we have an interesting sidelight on the trial in a letter to Southern (1793, June 17):

I have been confined to day with a headach and have yet received no account of what has passed in Westminster hall or with our lawyers & fear I shall not hear in time—...such quibbles as you mention form the very essence of the law, we had like to have been ruined because the patent instead J W of the City of Glasgow Mercht—says simply *J. Watt* the former being in the specification & in the act, which by the by has many such kind of errors. This mighty affair they say they have a remedy for, but there is something else which I cannot comprehend that they stick at, and keep us in a horrible suspense—

The verdict was for the plaintiffs, subject to the opinion of the court as to the validity of the patent. On May 16th, 1795, this special case came on for judgment, when the opinions of the Judges were equally divided. This led to a fresh trial and interminable delays; it was not till January 1799 that the final decision of the Judges in the Court of King's Bench was given; the judgment was affirmed, thus finally establishing the validity of the patent. Watt wrote to Boulton (1799, Jan. 25): "We have *won the cause* hollow. All the Judges have given their opinions carefully in our favour and have passed judgment. Some of them made better arguments in our favour than our own Counsel." James Watt, junior, let himself go with still greater vigour in a letter of the same date: "Send forth your Trumpeters and let it be proclaimed in Judah that the Great Nineveh has fallen; let the Land be cloathed in sackcloth and in ashes! Tell it in Gath, and speak it in the streets of Ascalon: Maberley and all his host are put to flight."

When the good news reached Soho there was tremendous excitement and rejoicing. To the plaintiffs the verdict was of the umost importance, for apart from the award of heavy damages and costs amounting to

D W

between five and six thousand pounds against the defendants, the swarm of infringers and the Cornish adventurers, who had sat on the fence refusing payment pending the result of the trial, could now be brought to book and made to pay up arrears. In this task both the junior partners were employed but it was well into the first decade of the nineteenth century before arrears were collected; after that the firm of Boulton and Watt was known no more in Cornwall. In legal circles the decision of the judges has constituted a precedent of importance in patent law.

It was on the eve not only of the expiration of the patent, which Watt now humorously alluded to as his "well tried friend", but also of the term of the original partnership and both partners must have felt it a happy climax. Boulton was now in his seventy-second and Watt in his sixty-fourth year. The great work of their life had been done. One would think they might have had a quiet time together as the anniversary drew near to talk over their triumphs, difficulties and failures; if so we have no record of it. For Watt the end of the partnership was a welcome surcease from toil, while for Boulton, as of old, it was an occasion to start a new enterprise; this time it was the improvement of coining machinery.

We mentioned at the beginning of the chapter Watt's purchase of land on Handsworth Heath; we must now mention further acquisitions of landed property. Mr and Mrs Watt had indulged in numerous tours to various parts of Great Britain. Of all places he was attracted most by the scenery of Wales. In 1798 he began to acquire land in the counties of Radnor and Brecon, particularly at Doldowlod on the Wye between Rhayader and Newbridge in the former county. The farmhouse there he transformed into a comfortable country house whither he resorted occasionally during the summer months and spent much of his time in making improvements on the property as was his wont. His son did much subsequently in the same direction. Doldowlod remains to-day the home of the Gibson Watt family, the collateral descendants of Watt.

When Watt said goodbye to Soho and all its works, he left it in capable hands and in a most flourishing condition. We have mentioned the capacity and energy of the junior partners but there was also the capable Southern looking after design and the drawing office, and last but not least the great Murdock who since he was back in Birmingham was finding effective outlets for his inventive faculties. Material advances were made in the construction and details of the engines, but into these we shall not enter as it is doubtful whether Watt had much hand in them; the hand was rather that of Murdock. Still it will help to mark definitely the position of affairs when Watt retired if we figure a typical example of the double-acting rotative engine then being supplied to customers. We have selected for representation from the Boulton and Watt Collection a mill engine made for Dixon, Greenhalgh & Welchman in 1802 (see Fig. 12). It will be noted how largely cast iron enters into the construction: the working beam, the connecting rod, the column under the beam and the

12-2

gearing are all of that material. The cylinder was 33⅓ in. diam. by 7½ ft. stroke.

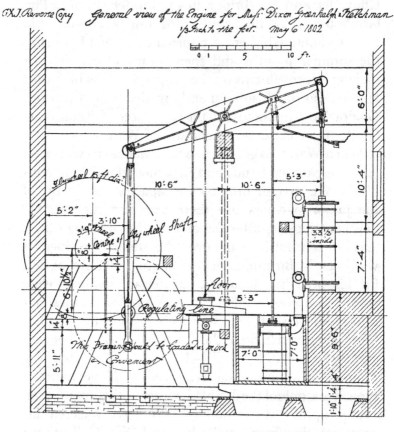

Fig. 12. Rotative engine for Dixon, Greenhalgh & Welchman, 1802

Boulton and Watt Collection. Courtesy of Birmingham Reference Library

Lastly, it is interesting to state what the steam-engine was capable of doing at the end of Watt's active career.

This is summed up for us succinctly in a letter from Boulton to young Watt (1796, Nov. 28) as follows:

One bushel (84 lbs.) of Newcastle or Swansey coal

(1) Will raise 30 million lbs. of water 1 foot high.
(2) Will grind & dress 10, or 11, or 12 bushels of wheat, according to the state of it.
(3) Will turn 1000 or more cotton spinning spindles per hour.
(4) Will roll & slit 4 cwt. of bar iron into small nailor's rods.
(5) Will do as much work per hour as ten horses.

CHAPTER VIII

LIFE IN RETIREMENT, 1800–1819

Travel. Deaths in family and among friends. Sculpturing machines.
Honours. Character. Death. Epilogue.

WATT at the age of sixty-four assumed the rôle of a country gentleman, as did at that time so many of those who had been enriched by industrial pursuits. He enjoyed the pleasures of the country in the summer time at Doldowlod and returned in autumn to the comforts of Heathfield and to the society of his friends in Birmingham.

He and his wife travelled considerably. In 1802, taking advantage of the short-lived peace of Amiens, they went up the Rhine to Frankfort, then to Strasbourg, and back via Paris to England. He visited Scotland frequently and he was staying there with Mr Gilbert Hamilton at Glenarbuck, near Dumbarton, when the whole country was alarmed by Napoleon's threatened invasion of England. There was a call to arms and everywhere bodies of volunteers were enrolled for home service; it was the beginning of the volunteer movement. Watt's idea of what he considered his duty in the matter is shown in the letter (1803, Aug. 31) to Boulton to have been £100 "to the fund at Lloyds" "and £10 to the Handsworth parochial fund". It does not look as if Watt was carried off his feet by excitement!

As one studies English industrial history of the

PLATE XV. JAMES WATT, *aet.* 77

From the oil painting by Sir Thomas Lawrence in the possession of
the Boulton family

period one is struck by the small attention apparently paid by the industrialists themselves to the momentous political changes that were going on around them; nevertheless, it was they who were providing the sinews of war that eventually "On that loud sabbath shook the spoiler down".

In 1805, Watt was again in Scotland, this time in the winter at Edinburgh, where Henry Brougham, afterwards Lord Chancellor, says "he was a constant attendant at our Friday club and in all our private circles; and was the life of them all". Watt was fond of Edinburgh and on a later visit, possibly in 1814 or perhaps 1817, he made the acquaintance of Sir Walter Scott. Their regard was mutual. The latter's panegyric on Watt is well known.

The wide extent of Watt's knowledge, his dry humour set off by a pronounced Scots accent that he never lost, the good stories he could tell, made him welcome in such company.

One of the penalties of old age is the loss of family and friends by death. The first gap in his own family since the death of his first wife was caused by the loss of his sister Margaret, who died in 1791; in 1794, to his great grief, his daughter Jessy died of consumption at the age of fifteen. The other child of his second marriage, Gregory, born in 1777, was a lad of considerable parts, handsome and of a generous nature. His father sent him to Glasgow College, where he had as fellow students Campbell and Thomson, the poets, and Francis, afterwards Lord, Jeffrey. They anticipated success for him at the Bar and possibly in Parliament, but his father intended him for commercial life and, as we have seen, made him a partner in Boulton, Watt & Sons. However, before he could take much part in the business, consumption supervened. Its oncoming is foreshadowed in an anxious letter (1797, Nov. 21) from his father to Thomas Wilson, asking him to find accommodation for the young man at Penzance.

During the time Gregory was there he lodged with Mrs Davy, mother of Humphry Davy. The latter was only a year older than Gregory and not unnaturally a friendship sprang up between the lads, leading to the young chemist being engaged as demonstrator to Watt's friend, the well-known physician Dr Thomas Beddoes. It may be recalled that the latter founded the Pneumatic Institution at Clifton, near Bristol, in 1798, and Davy's appointment there was his first step on the ladder that led him to fame. The Institution was established for the treatment of diseases of the respiratory organs by "factitious airs", i.e. artificially produced gases such as oxygen and carbon dioxide which were to be taken by inhalation. Beddoes was assisted in his venture by Watt, who had no doubt been drawn in because of his concern for the health of his own family. He collaborated with Beddoes in bringing out pneumatic apparatus, which were made by Boulton and Watt at Soho, and sold in some quantity. The results obtained by Beddoes were not up to expectation.

Gregory's health benefited by his stay in Cornwall, but after his return home the improvement was not maintained, although he derived some benefit from a tour on the Continent. His health steadily got worse and his anxious parents took him from place to place in the south in the vain hope of improvement. After trying Sidmouth, where the sea air did not suit him, he was removed to the neighbourhood of Exeter, only to die there a few days later on October 18th, in the 27th year of his age. It was a great blow to both parents, for Gregory had been their Benjamin.

The loss of old friends was another grief. Wedgwood had died in 1795, Black in 1799, Darwin in 1802, Priestley in 1803, and Robison died in 1805. Watt said of the latter: "He was a man of the clearest head and the most science of anybody I have ever known and his friendship to me ended only with his life after having continued for nearly half a century."

But he was to undergo a still greater loss in the person of his partner Boulton. The latter had suffered for a number of years from kidney trouble. In March he had a severe attack, from which he rallied, only to linger on through the summer; the end came peacefully on August 17th, 1809, at the age of eighty-one. Watt wrote to his son a letter (1809, Aug. 23) of condolence in which he said:

I am very much concerned to have to condole with you on the loss of your worthy Father.... We may lament our own loss, but we must consider from the other side his relief from the torturing pain he has so long endured & console ourselves with the remembrance of his virtues & eminent qualifications. Few men have had his ability & still fewer have exerted them as he has done & if to them we add his urbanity, his generosity & his affection to his friends, we shall make up a character rarely to be equalled. Such was the Friend we have lost & of whose affection we have reason to be proud, as you have to be the son of such a Father!

Watt was deeply sensible of the debt he owed to Boulton, and in an account of his life which he drew up expressed it in these words: "In respect to myself, I can with great sincerity say that he was a most affectionate and steady friend and patron, with whom during a close connection of thirty-five years, I have never had any serious difference."

The author has always felt it to be a matter for regret that a biography of this great man has never been written and further that Birmingham has never seen fit to erect an adequate memorial to his memory.

One of the fears of Watt in his old age was that his mental faculties, the exercise of which was his chief delight, might fail. To test them he took up again the study of German. He quickly re-acquired his early proficiency.

Not only did Watt retain to the end his mental faculties but, what is much less usual, he retained his inventive powers. Proof of this is furnished by his solution of a problem submitted to him in 1810 by the Directors of the Glasgow Water Works. The problem was to convey the filtered water across the river to the company's pumping station at Dalmarnock. Watt suggested, and supplied a drawing for, a flexible water main, on the analogy of the lobster's tail. The installa-

tion was successful and the Directors of the Water
Company were so delighted that, as Watt declined to
accept remuneration for his services, they insisted on his
acceptance of a service of silver plate of the value of one
hundred guineas.

A still better exemplification of Watt's inventive
powers in his old age is afforded by his work on the
sculpturing machine, and as this is so germane to the
subject-matter of the present volume we are fain to go
into it in considerable detail.

Watt's predilections, training and occupation in
early life had resulted in his becoming a skilful mechanic.
When he came to retire, it is not unreasonable to suppose
these early predilections should reassert themselves, and
that he should turn his thoughts to his tools in the garret
where they had been "dumped" when the family
removed from Regent's Place. Whatever the train of
events was, Watt made the garret his workshop and there
spent a great deal of his spare time. Nearly every man
loves to have, if he can, a den of his own where he can
work or idle as he chooses. If this were all in Watt's case,
no further remark would be called for except to say that
the contents of the workshop epitomise the career of
the man and reveal a good deal of character. But the
garret was more than this, for here he worked out the
last inventions of his life. By a happy sequence of
events the workshop was kept intact for over a century
and is now preserved as a memorial of the great
engineer.

It will assist the reader in understanding what follows

if he will refer to the illustration of the garret which forms the frontispiece of this volume.

The garret was on the south front of Heathfield, on the second floor over the kitchen, and was reached by the back staircase. It was a low room, some 20 ft. by 16½ ft., lit by a small skylight and a low sash window about 5 ft. wide overlooking the back yard. There was a chimney-breast where a fireplace had existed originally but it had been taken out and replaced, no doubt by Watt himself, by a stove for carrying out experiments that needed a somewhat high temperature. The stove was necessary also to warm the room, for, being so near the roof, it was very cold in winter; on the other hand in summer it was too hot for comfort. Our illustration (Fig. 13) shows the plan of the workshop as it existed at Heathfield, its arrangement and contents.

When Heathfield was about to be pulled down in 1924 Major J. M. Gibson Watt, the owner, decided to present the contents of the garret to the nation. The gift was gratefully accepted and when removing the contents to South Kensington, where they are now housed, an exact reconstruction of the garret was made; moreover, the door and frame, the window sash and frame and the flooring boards of the actual garret were incorporated in the reconstruction. To render the contents of the garret visible to the public, a plate glass window was inserted in the wall opposite the sash window and unobtrusive ceiling lighting was introduced. Every object was put into the place it had occupied at Heathfield, but as it is impossible to see a tithe of the objects, so many being

stored in drawers and boxes, a few of the more important
have been placed in a show-case outside the room. In

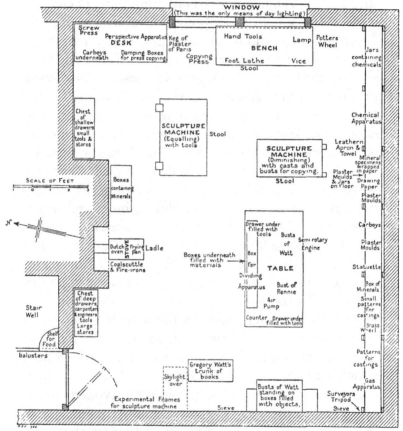

Fig. 13. Plan of Watt's workshop at Heathfield
Courtesy of the Science Museum

front of the little window is a work bench with a vice—
attached to the latter by string, so that it should not be
mislaid, is a centre punch—a little illustration of Watt's

method. On the bench is a bar lathe with treadle drive. Against the blank wall opposite the stove is rough shelving on which are stored jars containing chemicals, mineral specimens, plaster moulds, statuettes, patterns for castings, gas and other apparatus. On a nail in the shelving is a leathern apron, quite possibly the one Watt wore in the Glasgow College days. Beside it is a towel, dropping to pieces, and a clothes-brush whose bristles have fallen out; the presence of these objects is an indication that Watt, sensible man, removed the evidences of his work before descending to the living rooms, for Mrs Watt was particular. On the side wall, where the plate glass window is, are wooden patterns, frames belonging to the sculpture machines, busts and sieves. Then comes the door of the room, and if we look outside on what was the landing we shall see a bracket or shelf on the balusters. On this shelf, it is said, the maids placed food for Watt when he did not want to be disturbed. On the left of the door is a chest of drawers filled with carpenters' and engineers' tools, bulky stores and drawings. Next came the stove; upon this is a ladle used for melting lead; a dutch-oven and a frying pan, which lend colour to the tradition that Watt sometimes cooked a meal for himself in the garret to avoid breaking off to go downstairs. In the recess on the other side of the chimney-breast is a chest of small drawers, that dates back undoubtedly to the College days; each drawer is labelled in Watt's handwriting and filled with small tools and stores of every imaginable kind. On the left of the windows is a standing desk and near it are a number of his perspective machines of 1764 incomplete, damping boxes for press copying and a screw copying-press.

In the middle of the room on a table are busts of Watt himself, one of his friend Rennie, a semi-rotary engine, a counter, and an air pump. The outstanding objects are the two sculpturing machines, the fruits of Watt's invention and craftsmanship during the last years of his life, and these deserve a detailed account.

If we are to judge by the evidence of drawings found in the chest of drawers in the workshop, Watt must have begun on his first machine about 1804. By 1807, it must have been well advanced, for he had obtained from Turnerelli the sculptor "small busts of Socrates and Aristotle and a sleeping boy". Incidentally it is interesting to know that these figures are still in the workshop. In November 1808 Watt ordered some diamond-cutting pencils for what he calls his "Parallel Eidograph".

In March 1809 he ordered, besides other tools, "drills made in steel frames of peculiar construction to turn with great velocity and without a shake". By April 1809 he had "made considerable progress with the carving machine" and it seemed necessary to christen it, and he suggests a number of names derived from Greek words. Before the end of May he informed his friend Dr Patrick Wilson that he had now made the glyptic machine polyglyptic, as he could do two or more copies at once, but that it was "still far short of his ideas; *Ars longa vita brevis*". At the same time he told his son James that, since getting the drill frames (which he had obtained from a skilful mechanic named Green in Manchester with whom he had got into touch through Peter Ewart), he had "finished a large head of Locke in yellow wood and undercut, and a small head of Dr [Adam] Smith in ivory, which were both done on the iron platform; the former not so well as I expect to do, but very well".

So far we have found very little in the way of description of the machine, no doubt it underwent protean changes. We are, however, on fairly safe ground in believing that the basis of Watt's work was the *tour à medailles* or portrait lathe which in all likelihood he had seen in Paris in 1802. This belief is supported by the following letter (1809, July 7) to Dr Young, Professor of Greek at Glasgow:

There is a machine of the nature of a turning lathe which copies medals and other things in bas-relief: it is called in France tour à medailles, in England the likeness lathe. I have thought of some improvements on it which somewhat extend its uses (this is at present a secret which I do not wish to be spoken of).

It is a machine of this type capable of making larger or smaller copies only that is seen in the garret in the middle distance; but Watt evidently came to the conclusion that he would like to make a machine to produce copies of the same size as the original and to this he seems now to have given his attention. Throughout the year 1810 we have evidence that he received much friendly help in constructional details from William Murdock. Writing to Ewart (1810, May 3), Watt says:

You will see that instead of putting the drills in by screws, they are put in by a taper which I find easier fitted & is I believe easier of execution by him [i.e. Mr Green]. The drills are keyed from dropping out by a pin put through the spindle with a notch in the shank of the drill, but this notch should not be drilled out but marked & then cut by the file, so as to draw the drill in a little.

Socketing drills with a slight taper is now universal. Every mechanic will appreciate the little touch about filing the notch in the drill, so that the pin when driven in will draw the drill up into its socket; there speaks the craftsman.

Green turned out to be "incorrigible", however, and on May 27th Ewart was informed by Watt that he had "got our toolmaker at Birmingham to undertake the job".

We are fortunate in having, on the back of an old letter, notes of the successive stages, and a journal of the time taken, in making a bust of Sappho on the machine, and from this we gather that Watt out of 39 hours' total put in nine hours' work on one day and even snatched an hour on Sunday, so engrossing did he find this occupation!

Watt's final design of the drill spindles was the tapered socket with left-handed screwed shank, shown in the sketch (see Fig. 14), which he sent to Ewart, who replied that Green had undertaken to make the spindles and on June 6th forwarded two spindles and collars for drills with Green's receipt for £5 for making them. These are the spindles that are incorporated in the machine. There is in the workshop a quantity of drills, which we at the present day should call roughing and finishing milling cutters.

The treatment of the originals or patterns before being copied on the machine, required no little care, and we have in Watt's hand-writing, dated November 1st, 1810, several recipes. Even the composition of the "Cement to fasten the patterns to be cut to the tablets of the machine" was the subject of experiment.

We cannot, for lack of space, follow Watt's experimental work in detail, but by 1814 he had advanced so far that he contemplated taking out Letters Patent for the invention. He wrote to John Farey (1814, Aug. 10), the well-known patent agent, to say that he expected very soon to have the machine completed. By September he had drafted a remarkably concise and clear specification, a portion of which describing the proportional sculpturing machine only survives. What he would have claimed as novel we do not know. It is a pity the patent did not reach the Great Seal, for if nothing else it would have saved later inventors like Cheverton and Donkin from inventing it all over again. Drafting the specification seems to have been the culminating point of Watt's efforts.

There is a hiatus of two years before we hear anything further. The revival of interest might perhaps have been due to the fact that Watt had had his portrait bust executed in 1815 by Mr F. L. (afterwards Sir Francis) Chantrey, the most celebrated sculptor of his day.

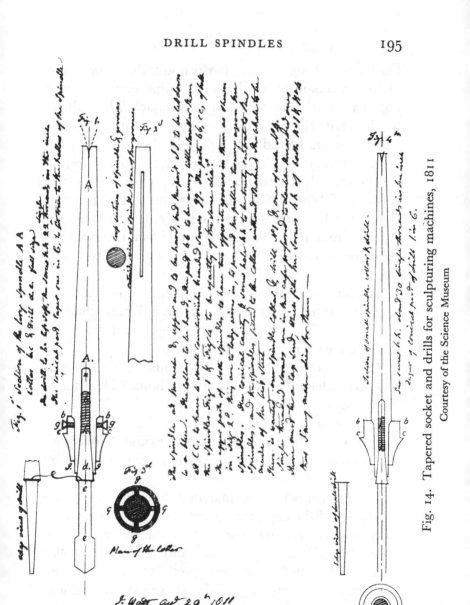

Fig. 14. Tapered socket and drills for sculpturing machines, 1811

Courtesy of the Science Museum

During the sittings we can hardly doubt that they would discuss mechanical methods of sculpturing the final object in marble from the artist's model. This subject would have been of peculiar interest to Chantrey, for, as is well known, he did not undertake the actual hewing of his statuary, but left it to assistants and further, he was frequently asked, as in Watt's case, to execute replicas of busts. While his assistants could produce replicas of the same size as the original, there were no means of making copies to scale, and it was no doubt the reducing sculpture machine in which Chantrey was interested.

The invitation to see the machine was sent, but when Chantrey next passed through Birmingham it so chanced that Watt was from home, and the great sculptor's wish remained unsatisfied.

The last reference that we have found to the carving work is in a letter (1818, May 26) to Thomas Thomson of Edinburgh.

From what has been said it is clear that the sculpturing machines underwent great changes and therefore all that we can do is to describe them as they were left by Watt.

In the proportional machine (see Pl. XVI) the original and the copy are mounted each on its carriage between centres; that means that the objects can be rotated and therefore a round and not merely a flat object like a medal can be copied. The carriages are connected by pantograph linkwork so that any reasonable proportion between the original and the copy can

PLATE XVI. WATT'S PROPORTIONAL
SCULPTURING MACHINE, *c.* 1818
Courtesy of the Science Museum

PLATE XVII. WATT'S EQUAL SCULPTURING
MACHINE, *c.* 1818

Courtesy of the Science Museum

be obtained. Both original and copy are commanded by a pivoted bar on which are the feeler and the cutting frame respectively; the latter is driven by a gut band over pulleys from the treadle, as is clear from the illustration. It will be realised that while the foot was working the treadle, one hand was manipulating the tool bar and the other the sliding carriage—strenuous enough exercise in all conscience for a septuagenarian!

The equal sculpturing machine (see Pl. XVII) consists of a treadle lathe bed supporting a tilting frame on which the original and the copy are mounted between centres, and these centres are connected by gearing so that they will revolve in step. The original and copy are commanded by jointed built-up tetrahedral frames. These frames are mounted on a slide on a vertical frame and can be traversed by a hand-operated ratchet gear. At the same centres as the original and copy are a feeler and a drill spindle respectively; the latter is driven by a gut band over guide pulleys from the treadle below. It will be realised that the operation of this machine was even more laborious than that of the proportional machine.

The closing years of Watt's life were pleasant ones, in striking contrast to those of most of his contemporary inventors. The fortune he had acquired enabled him to travel, mix with the outside world and indulge his simple tastes. He lived beyond the allotted span; honours came to him as the years went by—not posthumously as is too frequently the case. He was elected a Fellow of the Royal Society of Edinburgh in 1784 and in the

following year of that of London at the same time as
Boulton; their signatures appear together on the roll of
Members. In 1787 Watt was elected a member of the
Batavian Society, Rotterdam. In 1806 the University
of Glasgow honoured itself by conferring on him
honoris causa the degree of Doctor of Laws. In return for
this, he founded the Watt prize, very appropriately in
Natural Philosophy and Chemistry. The French
Academy elected him a Corresponding Member and in
1814 advanced him to the rank of Foreign Associate; as
one of eight such, this was a signal honour.

He could have enjoyed other distinctions, had he not
been of such a retiring nature, for he was asked to fill the
office of High Sheriff of Staffordshire and later the same
office in Radnorshire; in both cases he declined, giving
as his reason his advanced age. In those days it was rare
for the services of a scientific man and rarer still for an
engineer to have his services recognised by his sovereign,
but this nevertheless happened. During the premiership
of Lord Liverpool Watt was offered a baronetcy but
declined it with simple dignity. The coat of arms with the
motto "ingenio et labore", borne by the Watt family
and displayed on many of the monuments erected to
Watt's memory, was not assumed until 1826. We can
imagine with what irony Watt would have rejected the
idea of becoming armigerous!

In appearance Watt was above the medium height,
which his spare figure would have accentuated had he
not had a pronounced stoop of the shoulders. His eyes
were grey but in some lights looked blue. His com-

plexion was fresh-coloured, his hair turned white early in life. In repose his features bore the stamp of deep thought. He spoke in a deep and rather monotonous voice. He was not an early riser and required ten hours' sleep. He took snuff and liked a pipe of tobacco.

No one came into contact with Watt, even in his early days, without realising that he was a man out of the common. His character will have already been assessed by the reader, modest, self-depreciatory, cautious to a degree, unenterprising, a hater of business and disputes, but unremitting in application and above all fertile in invention.

Watt died at his home at Heathfield on August 25th, 1819, in the eighty-fourth year of his age; he had been in his usual health till the month before. He was buried in Handsworth Parish Church on September 2nd beside his friend Boulton. The funeral was carried out with the lugubrious ostentation of those days, despite the wish clearly expressed in his will that he might be "interred in the most private manner, without show or parade". By this will, dated July 7th, 1819, he left to his wife Ann £1400 per annum and Heathfield for life, and to his son James the residue of the estate, with which went all documents, drawings, tools, etc. The will was proved on October 13th for upwards of £60,000, a large sum for those days.

Since his death, Watt's fame has increased, and he is one of the few inventors whose merits have been recognised publicly, witness the statue by Chantrey set up by public subscription in Westminster Abbey in 1824. The well-known epitaph on the plinth of this monument from the pen of Lord Brougham has been acclaimed as the finest lapidary inscription in the English language. Not only is he remembered by statues and memorials, but Institutions, Colleges, Engineering Laboratories, Chairs, Libraries, scholarships, prizes and medals, to say nothing of docks and streets, have been named after him.

Looking back over Watt's career we must say that he was fortunate above all in his association with Boulton. Watt admitted that "without him the invention could never have been carried by me to the length it has been". We would go farther and say that while Watt's specification enrolled in the Patent Office would establish his claim to the important invention of the separate condenser, he would neither have brought his engine into general use, nor derived any reward for his invention, nor followed it up by those equally brilliant inventions connected with the rotative engine. The author has already argued that the extension granted by Parliament to the patent for the separate condenser was too long, and perhaps it was, but it must never be overlooked that but for this privilege the environment would not have been created wherein Watt worked out so many further improvements in the steam-engine. Glancing over the great changes that have taken place since Watt's day in every department of human activity, greater in extent than in any corresponding period of the world's history, we affirm that no one has made a greater individual contribution to these changes, for good or for ill, than has James Watt with his steam-engine.

BIBLIOGRAPHY

The ultimate source of nearly everything we know about James Watt is the documentary matter that, as stated in the Preface, has come down to us in such quantity. A brief account of this is desirable to shorten our notes of what has appeared about Watt in print. The MSS. are:

(i) Extensive documentary matter of every kind bearing on the whole life of Watt.
  In the family archives at Doldowlod.

(ii) A collection in the main complementary to (i) and no doubt withdrawn from it by Muirhead (see 2 below) when writing his biographical works.
  Presented by L. B. C. L. Muirhead, Esq., in 1921 to Birmingham Reference Library.

(iii) Correspondence of Watt with Dr Small and Matthew Boulton, 1768–1805.
  Part of the papers in the possession of the Boulton family at Tew Park, Oxon, until 1926 when the papers were entrusted for preservation to the Assay Office, Birmingham.

(iv) Records—Correspondence, drawings, accounts, etc., of the firm of Boulton & Watt and its successors, 1775 onwards. Of particular technical and economic interest.
  Presented to the City of Birmingham by George Tangye, Esq., 1911.
  Constitutes the Boulton and Watt Collection in Birmingham Reference Library.

(v) Correspondence of Boulton and Watt with Thomas Wilson, their agent in Cornwall, 1780–1803. About 1000 letters bound in 11 volumes 4to.
  Entrusted by Howard Fox, Esq., in 1909 to the custody of The Royal Cornwall Polytechnic Society at Falmouth.

(vi) Correspondence of Watt with John Southern, 1789–1815. About 100 letters bound in 1 volume 4to.
  Presented by R. B. Prosser, Esq., in 1915 to Birmingham Reference Library.

Consequently it is only necessary to cite the following works:

1. Williamson, George. *Memorials of the Lineage, Early Life, Education and Development of the Genius of James Watt.* 4to. 1856.
   Authoritative on Watt's early life; additional to the documentary matter above.

2. Muirhead, J. P., M.A. *Origin and Progress of the Mechanical Inventions of James Watt.* 3 vols. 8vo. 1854.
   Life of James Watt with Selections from His Correspondence. 8vo. 1859.
   Muirhead made use of (i), (ii) and (iii) above. Two standard works for all time.

3. Smiles, Samuel, M.D. *Lives of Boulton & Watt, principally from the Original Soho MSS.* In one or two vols. 8vo. 1865.
   Smiles made use of (iii) and 2 above. A most readable account of the development of the steam engine and Watt's share in it.

4. Dickinson, H. W. and Jenkins, R. *James Watt and the Steam Engine.* Watt Centenary Memorial Volume. 4to. 1927.
   Exhaustive study of the development of the steam engine under the hands of James Watt, using all material (i)–(vi). In this work a detailed bibliography of Watt is given, pp. 359–372, and to this we would refer readers who wish to go farther into the subject.

5. Roll, Erich. *An Early Experiment in Industrial Organisation, being a history of the firm of Boulton & Watt,* 1775–1805. 8vo. 1930.
   Based particularly on (iv). A valuable study from the economic standpoint.

INDEX

Academy, Watt associate of French, 198
Air pumps, 39, 42, 63, 97, 104
Albion Mill, 146–150
Anderson, Prof. John, 32
Apparatus for Boulton, 55
Arrêt de Conseil, 99, 100
Assay Office, Birmingham, ix, 201
Atmosphere, weight of, 6, 33
Ayr harbour, 74

Baronetcy offered to Watt, 198
Beams, experimental engine, 143, 144
Beddoes, Thomas, M.D., 184
Bedworth Colliery engine, 95–97
Bibliography, Watt, 201
Birmingham Reference Library, ix, 201
Black, Prof. Joseph, 24, 35, 185
Bleaching by chlorine, 151
Bloomfield Colliery engine, 90, 95, 101, 105
Boiler, haystack, 155; waggon, 156
Boring mill, 87, 167
Boulton, Matthew, F.R.S., 45; meets Watt, 47; his Soho Manufactory, 47; suggested share in patent, 50, 52, 62; Roebuck's debt to him, 82; partnership with Watt, 89, 178; confidence in Watt, 92; visit to Cornwall, 113; engages Murdock, 114; urges rotative engine, 124; introduces Southern, 131; Albion Mill, 146–150; describes flour mill governor, 153; his garden, 157; education of his son, 161; helps young Watt, 164; coining machinery, 178; death, 185; Watt's tribute to, 186, 200, *et passim*
Boulton, Matthew Robinson, 161, 162, 185

Boulton, Watt and Sons, 166
Bow, engine at, 95
Bridgewater Canal, 10, 44, 46
Broseley Ironworks blowing engine, 90
Brougham, Lord, 183
Brummagem wares, 48
Budge, John, 102, 105
Bull, Edward, 173, 175–177

Caledonian Canal, 85
Calonne, M. de, 151
Canals and navigations, 9, 44, 71–75, 78, 82, 84, 85, 112
Capital, raising, 113, 152
Carriage, steam, 139, 140
Cartsdyke, rise of, 11, 15
Cement, iron, 129, 130
Centrifugal governor, 153–155
Chacewater engine, 101, 102, 109, 110, 112, 113
Chaillot, engines at, 100
Chantrey, Sir Francis, 194
Chemistry, experimental, 30, 56; medical, 184
Chlorine, bleaching by, 151
Chronicle of the life and work of Watt, xv
Church and King riots, 159, 162, 177
Clyde estuary, deepening of, 11, 66, 67, 71, 74
Companies, City, 13
Compound engine, 133, 134, 173
Condenser engine, separate, 36–40, 46, 63–65, 97, 174, 175
Cooper, Thomas, M.D., 164
Copper trust, 152
Cornish folk, Watt's opinion of, 102, 112
Cornwall, engines in, 101–106, 109–115, 120, 169, 178
Cosgarne House, residence of Boulton and Watt, 124

CAMBRIDGE: PRINTED BY WALTER LEWIS, M.A., AT THE UNIVERSITY PRESS

Printed in the United States
By Bookmasters